ROUTLEDGE LIBRARY ED
LIBRARY AND INFORMATION SCIENCE

Volume 83

ROLE OF TRANSLATIONS IN SCI-TECH LIBRARIES

ROLE OF TRANSLATIONS IN SCI-TECH LIBRARIES

Edited by
ELLIS MOUNT

Routledge
Taylor & Francis Group

LONDON AND NEW YORK

First published in 1983 by The Haworth Press, Inc.

This edition first published in 2020
by Routledge
2 Park Square, Milton Park, Abingdon, Oxon OX14 4RN

and by Routledge
52 Vanderbilt Avenue, New York, NY 10017

Routledge is an imprint of the Taylor & Francis Group, an informa business

© 1983 The Haworth Press, Inc.

British Library Cataloguing in Publication Data
A catalogue record for this book is available from the British Library

ISBN: 978-0-367-34616-4 (Set)
ISBN: 978-0-429-34352-0 (Set) (ebk)
ISBN: 978-0-367-36453-3 (Volume 83) (hbk)
ISBN: 978-0-367-36454-0 (Volume 83) (pbk)
ISBN: 978-0-429-34622-4 (Volume 83) (ebk)

Publisher's Note
The publisher has gone to great lengths to ensure the quality of this reprint but points out that some imperfections in the original copies may be apparent.

Disclaimer
The publisher has made every effort to trace copyright holders and would welcome correspondence from those they have been unable to trace.

Role of Translations in Sci-Tech Libraries

Ellis Mount, Editor

The Haworth Press
New York

Role of Translations in Sci-Tech Libraries has also been published as *Science & Technology Libraries*, Volume 3, Number 2, Winter 1982.

The Haworth Press, Inc., 28 East 22 Street, New York, NY 10010

Library of Congress Cataloging in Publication Data
Main entry under title:

Role of translations in sci-tech libraries.

   Published also as: Science & technology libraries, vol. 3, no. 2, winter 1982.
   Includes bibliographical references.
   1. Science—Translating services—Addresses, essays, lectures. 2. Technology—Translating services—Addresses, essays, lectures. 3. Scientific libraries—Addresses, essays, lectures. 4. Technical libraries—Addresses, essays, lectures. I. Mount, Ellis.
Q124.R64   1983   418'.02'0245   82-23353
ISBN 0-86656-217-6

Role of Translations in Sci-Tech Libraries

Science & Technology Libraries
Volume 3, Number 2

Contents

EDITORIAL

The sci-tech literature of the world reflects the international nature of its sources—it is published in a wide variety of languages. Some scientists and engineers feel they may safely ignore materials written in languages other than their own native tongues, or perhaps they feel that obtaining translations is too much bother. Fortunately, these are not the only viewpoints maintained about the potential value of gaining awareness of and access to the many technical publications written in other languages.

There are many avenues available for obtaining translations of such materials, and sci-tech subjects constitute a major portion of the items being translated over the years. Some of the common means of obtaining translations include individual translations of books or journal articles, cover-to-cover translations of certain journals, and central collections of translations. Government agencies, business firms, individuals working alone, or professional societies are among those groups which serve as translation sources. This issue is devoted to a review of some of these sources and their activities, as well as ways in which one can become aware of what has been translated and obtain copies.

The first paper, by Suzanne Fedunok, reviews the various printed and online sources of obtaining copies of translations, or of commissioning a translation to be prepared. The scope and services offered by the National Translations Center is covered in an article by Ildiko D. Nowak. The role of commercial translating firms is described in Robert L. Draper's paper, followed by a paper by Fred Klein in which he explains the extent and nature of translators' work in the United States over the past few years. The way in which an academic institution can use students to handle technical translations is covered in the paper by Victor Hertz. The treatment given translations by *Engineering Index* (and other Ei products) is the subject of the paper by Zoran Nedic and Barbara S. McCoy.

The special paper for this issue, written by Wilda B. Newman, deals with the controversy over the role of private enterprise in the handling of government information.

Among the listings for our regular sections readers may note that the editor of *Sci-Tech in Review* is now Suzanne Fedunok, Mathematics/Physics Librarian at Columbia University. She replaces Susan Pasquariella, who had to resign the post due to the press of business.

This issue includes comments on patent searching developments at the New York Public Library in the *Sci-Tech Notes* section.

Ellis Mount

EDITOR'S NOTE: Readers interested in further information on technical translations may want to see the Fourth quarter 1982 issue of *Technical Communication*, published by the Society for Technical Communication at 815 Fifteenth St. NW, Washington, DC 20005. The entire issue is devoted to translations.

PRINTED AND ONLINE SOURCES
FOR TECHNICAL TRANSLATIONS

Suzanne Fedunok

ABSTRACT. Following a discussion of problems connected with the major sources of technical translations—cover-to-cover journals, unpublished or ad hoc translations, databases, and commissioning a translation from a translator or national translation center—these sources are reviewed in more detail. An annotated bibliography of the most recent sources of information on the subject is added as an appendix.

In a recent paper B.J. Birch of the British Library Lending Division makes the point that "any English-speaking scientist is very naive to believe that he misses nothing if he does not bother with references in foreign languages."[1] While approximately fifty percent of all scientific publications are in English, supporting the assertion that English is the universal language of science these days, it would indeed be a great mistake to ignore the scientific literature in foreign languages.

Not only has the trend toward national awareness caused a shift away from English, but also increased industrialization and national prosperity are causing useful information to appear in more and more languages. Online access to multilingual databases is also contributing to the need for translations, a need which is growing, not shrinking, as scientists attempt to keep up with the burgeoning scientific literature in their fields. In 1979 the British Library Lending Division dealt with over 18,000 requests for translations.[2]

A survey conducted by the BLLD in 1978 on foreign language problems facing the research worker showed a decline in the proportion of scientists and technologists able to deal with foreign languages since an initial BLLD study in 1965.[3] As fewer and fewer scientists master foreign languages, the translation problem is compounded. Even French and German are not studied in as great depth nowadays, and Russian, which comes second to English in the proportion of literature published, has severely declined. This means that access not only to important new research, but also to important papers

Suzanne Fedunok is Mathematics/Physics Librarian for Columbia University Libraries, New York, NY 10027. She received the AB degree from Bryn Mawr College and the AM and MLS degrees from the University of Michigan.

3

of the past, is increasingly restricted because translations of the older papers were simply not made at the time they appeared. Researchers could read more languages then. Language problems put pressure on the science and technology librarian to provide adequate access to the literature of translations, a literature which, as we shall see, is not yet under very rigorous control.

D. van Bergeijk, the Director of the International Translations Centre in Delft (Netherlands), refers to a "double barrier" in dealing with the translations literature. There is the language barrier, which to his mind involves identifying relevant citations in languages not familiar to the reader, (and which is bound up with the need for multilingual guides and indexing) and there is the document delivery barrier.[4]

There are of course many aspects to this document delivery problem, the most obvious being the identification of translation documents to deliver. One recent study of International Translations Centre requests found that of 1764 translations requested, 1351 were not found, and as a rule of thumb the ITC reports that only 25% of their searches are successful.[5]

A second obvious aspect of this problem is the long time lag between the appearance of the original paper and its subsequent translation, provided a translation is ever made. It is not unusual in science libraries to encounter readers aware of the existence of a key paper in a foreign language long before that paper has been translated. This tends to support the contention that the communication network provided by the "invisible college" of scientific researchers in some way obviates the need for translations. This is not the case. Often a reader is going on hearsay only, or has seen the paper cited in an article or an index, and the reader does not possess or expect to possess a preprint or reprint of the article. The reader thinks he urgently needs to see the entire paper in a language he can read. This will always pose problems for the library staff, since the average lag time for translations to appear, as established by the ITC, is 18–24 months.[6]

J. D. Anderson of the Rutgers University School of Library Science suggests one possible solution to this time lag problem: the prior translation of articles that can be predicted by citation analysis to have a wide impact and their publication in a "selected translations" format.[7] This proposal is controversial and has not yet been adopted by publishers.

When a translation is not expected in one of the cover-to-cover or selected translations journals, there are at least two possible courses of action, both of which involve problems. An experiment at the BLLD found that over half of requestors who were given English summaries of articles for which they had requested full translations indicated that they no longer needed to see the full translation.[8] Many papers do include English summaries, and another obvious source is the indexing and abstracting journals. The problem is to obtain a copy of the original, which is likely to be from an obscure journal, and to provide an English summary if one is lacking. Should the English summary prove inadequate, the next problem might be to provide a full translation of the paper.

In the case where the original is not available "in-house," the usual sources for interlibrary loan should be tried of course, but this usually involves unacceptable time lags. Informal networks either of the reader's or of the library might be better used at this stage, or the decision made to go directly to the National Translations Center for assistance.

The preceding few paragraphs are intended as a quick sketch of the major problems involved with the document delivery barrier to access to translations. What follows will be a more detailed look at the various ways of delivering translations to the anxious reader. An annotated list of major sources follows the text of the paper.

Published Translations

Bibliographic control of published translations is usually achieved in the same way as for the original literature. Cover-to-cover translations and the more recent "selected translations" are covered by the major abstracting and indexing services, which usually list both the original title and the title of the translation journal in their source lists. Perhaps the most comprehensive treatment of translations, in terms of bibliographic control at least, is that provided by the American Mathematical Society. *Mathematical Reviews*, which has a policy of reviewing from the original, will add the following sentence to its reviews as it applies: "This article has appeared in English translation" follows the list of source journals in the annual index volume, and references are made back and forth from original to translation. The heading of each paper reviewed in the *Reviews* also includes information about foreign language summaries in the text of the paper. Translations from all languages are commonly indexed in the *Reviews*, which refers the reader to the original version.

The International Translations Centre finds that they receive a considerable number of translation requests which refer to papers already published in cover-to-cover translation journals and which should be readily accessible. In their study of 1764 requests already mentioned above, 192 of these were published in cover-to-cover or selected translations journals.[9]

There are 1173 publications listed in the 1974 edition of *Translations*

Journals, of which about half represent cover-to-cover journals. (More complete bibliographic information to sources mentioned in the text follows in the List of Sources appendix.) About 250 of this number may be considered a ''core'' collection and are subscribed to by the BLLD.

No science librarian needs to be reminded how expensive are subscriptions to translations journals. Their number is limited, and it may be shrinking. Few libraries subscribe to even the 250 core translations in science and technology, so that guides to these journals are essential. The standard sources are: *A guide to scientific and technical journals in translation* (SLA, 1972), and *Journals in translation* (BLLD/ITC, 1978).

One difficulty that might account for the number of published translations not found in the ITC study could be mistakes in transliteration from non-Latin alphabets, since standardization in this area has hardly been universally accepted. Most published indexes and journals of translations will at least mention what transliteration system they use, and some provide conversion tables.

A more serious problem with cover-to-cover translations, and one already alluded to, is the lag time in publication. Often one late translation will delay an entire issue, and lags often exceed two years. One way to circumvent this problem is to contact the editorial offices of the translation journal and request a manuscript or preliminary copy of the translation. Provided such a translation is available, several large publishing houses specializing in translations will oblige, although sometimes for a fee. Consultant's Bureau and Allerton Press are two examples, and their fees run around $8 per article.

Unpublished Translations

Another large category of translations is commonly referred to as ''unpublished'' or ''ad hoc,'' although access to this literature does exist. These publications are generally issued by government agencies or research laboratories, by universities and business firms. They include translations of journal articles, patents, specifications, technical reports and preprints.

A significant number of the ITC requests received in the study already mentioned were found in the report literature (48 out of 413 located). As with the cover-to-cover journals, the control of report literature is maintained by standard sources with which every science librarian should be at least somewhat familiar. Information about translations is imbedded in these indexes and databases and must be sought as any other technical report or patent is sought. The *U.S. Government Reports Announcements and Index* and the USGPO *Monthly Catalog of United States Government Publications* are the standard U.S. sources, and they are fortunately available online.

Another agency which does a lot of translating, often in unexpectedly broad areas, is NASA, which publishes the *Scientific and Technical Aerospace Reports (STAR)* and *International Aerospace Abstracts* (the latter through a contract with the American Institute of Aeronautics and Astronautics). Both of these indexes are also available through the RECON database. NASA's *STAR* has a special reports category devoted to translations, the TT-F series; it seems to be the only agency doing this.

Another government agency which bears mention in this context is the U.S. Department of Commerce's Joint Publications Research Service (JPRS), which publishes translations of items selected from foreign-language sources by various government agencies and departments. There are finding aids published by the JPRS to these ''ad hoc'' translations, some of which are also collated into series by geographic area.

Certain industries, notably the metals industry, have established their own clearinghouses and guides to translations in their subject areas. For example, in 1957 the British Industrial and Scientific International Translations Service (BISITS) was created by the leading British steel producers. Now expanded to include both ferrous and non-ferrous metals, and under the umbrella of the Metals Information Office of the Metals Society (British) in London, it prepares, indexes, and distributes over 1250 translations a year. A bi-weekly list of its translations is available, with prices included, from BISITS, the Metals Society, 1 Carlton House Terrace, London, SW1 5DB, England.

The ASM Translation Service, established in 1975, by the American Society for Metals, acts as a central clearinghouse for the world literature of translations into English on the subject. The office receives the BISITS translations, as well as those from a number of other agencies: Henry Brutcher Metallurgical Translations, World Aluminum Abstracts, V.E. Riecansky Translations, Steel Castings Research and Trade Association Translations, National Institute for Metallurgy Translations (South Africa), BCIRA Translations and R. H. Chandler, Ltd. These translations are fed into a computer-stored database, which now includes over 20,000 items. Access to this database is available through *Translations Index: A Quarterly Source and Author Index to the Available Translations into English of Technical Papers in Metals and Materials* ($80/yr. and available from the American Society for Metals Translation Service, Metals Park, Ohio 44073). The cost of individual translations listed is about $20, and the articles translated seem to average approximately 5 pages in length.

A number of national clearinghouses for translation information have attempted to control this ''ad hoc'' literature, and are sharing information gathered inside their borders. The two most important sources for English speaking readers are the *Translations Register-Index* of the U.S. National Translations Center and the *World Transindex* of the International Translations Centre in Delft.

Translation Services

Probably the most expensive solution to the translation problem is to have a translation "custom-made" for a researcher. Although the rates charged by the National Translations Center are not unreasonable, and will provide wider access to the translation through its depository program, there are other potentially cheaper and faster ways to proceed, which should be looked into first. The cheapest way to get a translation done is to find someone on the staff who knows the language and would be able to make at least a rough English summary, remembering that in most cases this will prove sufficient. In universities and large firms this might not prove much of a problem. Often the language departments of local universities can put one in touch with qualified foreign students to do translations, and many colleges and universities run translation centers that charge reasonable rates. The fee at Columbia University runs from 4–14¢ per word or character, depending on the difficulty of the language, technical vocabulary, and length of the item. Some large firms even have staffs of permanent translators, or have contracts with free-lance translators or with translation agencies. The library staffs of large firms may keep files of competent translators. Then there is the telephone book; there are 127 entries under translators in the Manhattan directory. Finally there are published directories of translators, such as that of S. Congrat-Butlar (*Translation and translators: an international index*, Bowker, 1979).

Commercial translation agencies assign the work to the most competent person on the staff, and are also in theory responsible for the accuracy of the translation. The rates may be higher than those of a university service or a free-lance translator, but it might be worthwhile if the subject matter is quite specialized and if a strict deadline must be met. In general, free-lance translations will need more background information and more assistance in their work, as they cannot usually afford to be very specialized. One should try to have the translator translate into the mother tongue.

Translation Centers

The national translation centers have accepted the responsibility to provide original translations on a charge-back basis, provided that a copy of the translation be put on deposit at the center and made available to others later.

As B. J. Birch puts it, "It is a matter of common sense that a copy of every translation should be filed somewhere for permanent reference."[10]

In the United Kingdom, the British Library Lending Division has taken on the job of serving as the national translation center there. It will commission from an outside panel of experts the translation of any article in any language, provided no prior translation exists. The cost of the service there is about £ 1 per 1000 words or characters, and the average time lag is 4–5 weeks. The translations are then "notified" to the public through the monthly *BLLD Announcements Bulletin*.

Through a reciprocal arrangement the BLLD translations are made available to readers in the United States through the National Translations Center, which operates in much the same way. (See the paper elsewhere in this journal.)

The NTC was established in 1953 by the Special Libraries Association and is now part of the John Crerar Library, 325 W. 33rd Street, Chicago, IL 60616 (telephone 312-225-2526; TWX 910/221-5131). The collection holds over 325,000 translations and has location information on many others. It was created to be both a depository and an information source for translations into English of the world literature of the natural, physical, medical and social sciences. It will undertake to do bibliographic searching to determine the availability of translations and to determine prices of those available from commercial sources. The NTC has an exchange agreement not only with the BLLD, but also with the ITC in Delft.

Similar national translation centers exist in France, Germany and the Soviet Union, all of which are cooperating with the International Translations Centre in Holland.

The International Translations Centre maintains close contacts with all sources of information on translations, especially the national centers, with the object of "encouraging and promoting the use of literature published in less accessible languages of interest to science and industry, and to promote international cooperation in this field."[11] Notifications of translations are published in the *World Transindex*, searching for published translations is done on request, and documentation is collected on translations and translating. Like the National Translations Center, the ITC maintains its information in computer storage, but it prefers to consider itself as a network connecting the various national centers, and not as the first and primary source for translations. As its director D.van Bergeijk points out, the job of documenting and controlling the world's translation literature is much too great to be dealt with by one single institute.[12]

Databases

Apart from the online databases mentioned earlier that provide access to abstracting and indexing journals that also cover translations, there exist the computer files of the NTC and the ITC, which deal exclusively with translation literature. The ITC has announced that the *World Transindex* database will be available online. The database is produced at the CNRS in France and is accessible via ESA/IRS (European Space Agency Frascati). Information on how to use the database was published in a technical report in 1979; at the end of that year there were 49,707 records on file.[13] There do not seem to be similar plans for the NTC, but as the records of the ITC in theory include everything notified to the NTC, this might not prove a handicap.

Conclusion

Despite the promising developments and the increasingly better control achieved over translations these days, they still present many problems. The quality of indexing varies from product to product, and it still seems to be impossible to predict where and when a translation will appear, if it is not published in a cover-to-cover journal. The advent of online databases has helped greatly to facilitate access to this literature, but the world translations index is still a dream of the future. Few users realize, for instance, that the *World Transindex* does not include many "interwestern" languages, and that their inclusion would add over 16,000 translations to the files per year. Access to Japanese and Chinese literature is still inadequate, and there is still no national collection of translations from German. Much remains to be done.

Nevertheless, it remains true that few library users are aware of those sources that do exist to track down translations, and they are often stopped cold in their search. In mastering this tricky and elusive literature the science and technology librarian is indeed providing an invaluable service to researchers and technologists, one that they do not seem able to undertake on their own.

REFERENCE NOTES

1. Birch, B. J. Tracking down translations. *Aslib Proceedings.* 31(11): 500–511; 1979 November.

2. van Bergeijk, D. Developments in the bibliographic control of translations with particular reference to the activities of the International Translations Centre. *Interlending Review.* 8(4): 128–131; 1980 October.

3. Ellen, Sandra R. Survey of foreign language problems facing the research worker. *Interlending Review*. 7(2): 31–41; 1979 April.

4. van Bergeijk, op. cit., p. 128.

5. van Bergeijk, op. cit., p. 129.

6. van Bergeijk, D.; Risseeuw, M. Organizational profile: the International Translations Centre; the language barrier in the dissemination of scientific information and the role of the ITC. *Journal of Information Science* 2(1): 37–42; 1980 August.

7. Anderson, James D. Ad hoc and selective translations of scientific and technical journal articles: their characteristics and possible predictability. *Journal of the American Society for Information Science*. 29(3): 130–135; 1978 May.

8. Wood, D. N. The foreign language problem facing scientists and technologists in the United Kingdom; report of a recent survey. *Journal of Documentation*. 23(2): 117–130; 1967.

9. van Bergeijk, op. cit., p. 129.

10. Birch, op. cit., p. 508.

11. van Bergeijk, op. cit., p. 129.

12. van Bergeijk and Risseeuw, op. cit., p. 41.

13. Pelissier, D.; Mallet, M. Acces en conversationnel au fichier *World Transindex*. Rapport TRITA-LIB 4063, Stockholm. Papers in Library and Information Science, August, 1979. Royal Institute of Technology Library, Stockholm, Sweden.

APPENDIX

List of Sources

Cover-to-cover and Selected Translation Journals

Journals in translation. 2d ed. British Library Lending Division, Boston Spa, West Yorkshire, LS23 7BQ, U.K. and International Translations Centre, 101, Doelenstraat, 2611 NS Delft, the Netherlands. 1978. 181 p. ISBN 0-85350-171-8. £ 18.

> Contains an annotated alphabetical list of 982 journals with order information; keyword in context index of main titles; original title list; key to publishing and distributing agents. There are appendices on Soviet patents, the *Doklady* and the Institute of Electrical Engineers in Japan. A comfiche version of this source is available from: Instituto de Informacion en Ceincia y Tecnologia, Madrid, Spain. Supersedes *Translations journals—annex to the World Index.*

Himmelsbach, Carl J.; Boyd, Grace E. *A guide to scientific and technical journals in translation.* 2d ed. New York: Special Libraries Association, 1972.

> This source evaluates the translations and gives full bibliographic details on each. Out of date, but still useful.

Unpublished/ad hoc Translations

Translations Register-index. Compiled by the National Translations Center, John Crerar Library, 325 W 33rd Street, Chicago, IL 60616. 1967- . monthly. $78/yr USA; $88/yr foreign.

> Published in two sections: the Register of newly received translations arranged under the 22 COSATI categories; and the Index, a computer-produced journal citation index arranged by original journal article. Supersedes *Technical translations* (U.S. Dept. of Commerce, vols. 1–18, 1959–1967).

World Transindex. International Translations Centre, 101, Doelenstraat, 2611 NS Delft, the Netherlands. 1978- . monthly. Dfl 725 or $ equivalent.

Supersedes the *List of translations notified to ETC*, and represents a merger of *Transatom Bulletin and the Bulletin des Tranductions*. Centralizes the translation announcements of the ITC, the Commission of the European Communities and the CNRS. Presents about 2000 references a month, including patents and standards, indexed under COSATI subjects, sources and authors. Replaces the *World Index of Scientific Translations*. The annual cumulation cumulates the Index only; the other eleven issues must be kept, since they contain bibliographic information. Lists translations from Eastern European languages only (until 1978), now also includes translations into French from all languages and (from 1979) translations into Spanish. Only index source to translations published as part of the U.S. Joint Publications Research Service.

BLLD Announcement Bulletin. British Library Lending Division, Boston Spa, Wetherby, West Yorkshire, LS23 7BQ, U.K. 1973- . monthly. price n.a.

Lists British report literature and translations from all languages produced by British government organizations, industry, universities and learned societies. All items listed are held by the BLLD and are available in photocopy. Entries are arranged in subject categories, but lack author, source, or report number indexes. A list of sources producing the translations follows the subject section. Supersedes *NLL Translations Bulletin* and continues *BLL Announcement Bulletin*.

Chillag, J. P. Translations and their guides. *NLL Review*. 1(2): 46–53; 1971.

A rather dated survey of lists of and indexes to translations. It concentrates on the U.S. and the U.K. and mentions the cumulated card indexes of Aslib and the BLLD. A table gives data on 31 translations announcement bulletins.

Aslib index of unpublished translations. Aslib, London, 19   n.p.

Although Aslib holds no translations, it does maintain a card index of over 500,000 entries compiled from over 400 organizations in the UK and abroad, including the NTC holdings. The information is available to corporate members of Aslib, including those located overseas. Inquiries may be made by telephone, telex or letter.

Translators and Translation Services

Congrat-Butlar, Stefan, compiler and editor. *Translations and translators: an international directory and guide*. New York: Bowker, 1979. 241 p. $40.

Includes a ''Register of translators and interpreters,'' subdivided by agencies and subjects, including scientific and technical. There is a long bibliography at the end.

THE NATIONAL TRANSLATIONS CENTER: ITS DEVELOPMENT, SCOPE OF OPERATION AND PLANS FOR THE FUTURE

Ildiko D. Nowak

ABSTRACT. A review is presented of the developments leading to the founding of the National Translations Center. The sudden surge of multi-lingual scientific and technical literature in the post World War II years, and especially during the Sputnik era, brought into focus the handicap of the mono-lingual US professional community. For the past 30 years, the National Translations Center has been serving as an international depository and information source for English scientific and technical translations available from any known source in the US and other English-speaking countries. The scope of operation and services of the Center are explained in some detail. Various translation accession lists published over the years are discussed, as is the Center's monthly index. Future plans and goals are presented.

Background and History

Latest developments and research in foreign countries contributed to the world-wide advancements in science and technology resulting in the improvement of life for mankind. However, this same improvement introduces complications. A great portion of modern research is carried out in non-English-speaking countries. Published literature is the most effective means of exchanging knowledge; and translations are the only solution in providing access to multi-lingual information resources. Since its founding in 1953, the National Translations Center has been a pioneer organization working toward the goal of coordinating translations efforts and creating a central source which would be accessible to all and be to the mutual benefit of originators and users of translations alike. For the last 30 years the Center has served this purpose.

In the years following World War II, the US professional community became truly aware of the vast amount of scientific and technical research carried out in other countries. The post-war release of German reports and

Ildiko D. Nowak is Chief, National Translations Center, 35 West 33rd Street, Chicago, IL 60616.

other documents made the mono-lingual American scientists acutely conscious of their handicap. Translations became the only solution to the problem. Large numbers of foreign reports were being translated, in many cases simultaneously by different organizations. A group of concerned members of the Special Libraries Association's Science and Technology Division became worried about the waste of time, money and effort caused by duplication of translations. Means of reducing the duplication and organizing the translations activities were debated. At first, voluntary registration and exchange of translations was attempted. However, it soon became evident that this task was too large to be handled by volunteers. Furthermore, identification of organization originating the translation became a serious drawback. When identified as the originator of a specific translation, many organizations, especially industrial research centers, felt that they were revealing their field of interest and research. The need for establishing a ''clearinghouse'' for translations became apparent. A neutral, preferable non-profit, location should be found. The problem was somewhat alleviated when in 1953 The John Crerar Library was selected as the site for the Special Libraries Association's Translations Pool, where translations were to be collected, processed, announced and copies supplied upon request. The SLA Translations Pool was operated by The John Crerar Library under contract with the Special Libraries Association. The initial collection of some 1,500 translations became the nucleus of the SLA Translations Pool, later named the SLA Translations Center.

The first step in organizing the translations was the publication in 1953 of a list of translations in the original collection under the title *SLA List of Translations*, followed by a *Supplement* in 1954. By this time a substantial number of industrial, governmental, academic and research libraries were depositing their translations with the SLA Translations Pool. The deposits were made with the stipulation that the translations would be processed, cataloged and announced, and provisions would be made to supply copies of the translations upon request. The collection was growing at an average rate of 180 new translations per month. Persons in need of translations were astonished when they learned how expensive custom translations were and how difficult it was to locate existing translations. The need for a regularly published list of new additions to the SLA Translations Pool was apparent. With the help of initial grants from the National Science Foundation, the National Institutes of Health and the American Iron and Steel Institute, publication of a regular translations accession list became a reality. The first volume of the new accession list, *Translations Monthly*, was published in 1955.

While the SLA Translations Pool was engaged in collecting and processing translations from Western and Oriental languages, the Scientific Translations Center at the Library of Congress was engaged in similar activities covering translations from Russian only. Translations collected at or reported to the Scientific Translations Center were announced in the *Bibliography of Translations from Russian Scientific and Technical Literature*, 1953–1956.

In 1957 the SLA Translations Pool changed its name to SLA Translations Center and expanded its activities to cover not only translations deposited with the Center but also those available from commercial translating agencies and professional societies. The Center's efforts and contributions in making translations accessible were acknowledged when, in 1957, the Library of Congress transferred its entire collection of translations into Russian. The Center assumed the services formerly provided by the Scientific Translations Center.

The launching of Sputnik I in October of 1957 suddenly highlighted the importance of research in the Soviet Union. This was followed by rapid development in modern technology in Germany, latest electronic advancements in Japan, as well as intensified research in all countries around the world. The natural consequence of fast growing world-wide research was a publication explosion. Almost 50% of the scientific and technical literature was published in languages other than English. Industrial research facilities and governmental agencies increased their translation activities considerably. The number of new translations deposited with the SLA Translations Center more than doubled, the *Translations Monthly* expanded its coverage to encompass translations from all languages, including Russian, and from all fields of theoretical and applied sciences.

In recognition of the services extended by the SLA Translations Center, the National Science Foundation provided grant support to help offset the cost of operating the Center. To fully serve its users, the Center has established exchange agreements with national groups and professional societies around the world, by means of which translations are deposited with, or reported to, the Center. Among these are the British Library Lending Division in Boston Spa, Great Britain, the Canada Institute for Scientific and Technical Information of the National Research Council of Canada, Commonwealth Scientific and Industrial Research Organization of Australia, Council for Scientific and Industrial Research of South Africa, the International Translations Centre in Delft, the Netherlands, to name a few. Continuing contributions of translations were actively solicited from industrial and other special libraries, governmental agencies, academic institutions, professional societies and commercial translating agencies, in an effort to cover all known sources of translations. To assure an anonymous input and to protect the depositors' field of interest, all indication of the donor or translator is obliterated from the translations. Once a translation is processed and part of the Center's collection, the identity of the depositor is no longer known.

In 1959, the increased US Government interest in translations, mainly from Russian, led to a joint project in which the Office of Technical Services (OTS), now the National Technical Information Service (NTIS), collected and cataloged all government-produced translations, while the SLA Translations Center was concerned with translations produced by the private sector. OTS published *Technical Translations* (1959–1967), a combined index of translations processed by both OTS and the SLA Translations Center. The relationship lasted through 1966. At the end of 1967, *Technical Translations* ceased publication.

Partial funding from the National Science Foundation continued until mid-1973, enabling the Center to continue its function as the principal US depository for translations. The Center carried on its responsibility for translations produced by the private sector. Furthermore, those government agencies aware of their obligation to provide immediate access to their translations choose to deposit these with the Center. In 1967 *Translations Register-Index (TR-I)*, the Center's official publication, became the announcement medium for all translations into English, regardless of source from which these may be obtained. *TR-I*, published monthly, is composed of two parts; the Register section listing full bibliographic data for selected recent translations, and the Index section, a computer-produced citation index to all new translations deposited or reported during the preceding month. Each *TR-I* issue presently lists 1,500 newly available translations. The importance of this publication is demonstrated by its world-wide distributions.

By the mid-1960s the Center's translations information master file contained records of over 150,000 individual translations available from several sources in the US and other English-speaking countries. Many lists of translations available from numerous sources have been issued by various organizations throughout the years. The existence of so many lists caused serious complications in determining not only whether a given item exists in translation, but also its actual availability. The feasibility of consolidation of all existing translations listings into a single availability index was studied by the SLA Translations Activites Committee. Specific proposals for compilation of all translations information data were prepared. These efforts were rewarded in 1966 with a grant to the Special Libraries Association from the National Science Foundation for publication of the consolidated index. A specialized computer program was designed for storage of translations information and its retrieval for indexing. Work on the index began at the SLA Translations Center and was completed by the end of 1968. The final product, the *Consolidated Index of Translations into English (CITE)*, published by the Special Libraries Association in 1969, contains records of some 150,000 translations into English available from any known source in the US and other English-speaking countries. Included are selective translation journals and special collections of translations. Journals translated cover-to-cover (published in English edition) are identified. All entries are by citation to the publication where the original foreign paper was published. The computer program originally written for CITE was used for the Index sections of *Translations Register-Index*. While the supply lasts, copies of CITE may be purchased from the National Translations Center at the price of $19.50.

Current Status

In 1968 the SLA Translations Center became independent of the Special Libraries Association and assumed its present name. The National Translations Center is now under the administration of The John Crerar Library. In 1973 support from the National Science Foundation was phased out, and the Center was placed on a self-supporting basis. The Center's only income was derived from subscriptions to *Translations Register-Index* and from a service fee charge when a translation is supplied. The lack of operating funds resulted in a drastic reduction of staff, leaving the Center with a staff of only four. In spite of its financial difficulties, the Center continues to fulfill its role as an international depository and information source for translations. The Center operates as a unit of The John Crerar Library concentrating all its efforts on translations. The idea of sharing translation efforts has caught on. Through the cooperation of organizations from the private sector, as well as the government, the collection now numbers over 325,000 individual translations. In addition to its own collection, the Center maintains up to date records of translations reported regularly by over 100 sources around the world. The master file contains data on some 800,000 existing translations from all fields of theoretical and applied sciences. On the average 1,500 new translations data are added each month. The languages most frequently translated are Russian, German, Japanese, French, Italian (in that order), followed by other European and Slavic languages. Recent additions to the Center's translations information data bank show an increase of translations from Chinese.

When translations are received at the Center, their receipt is immediately recorded and all identification of the depositor is obliterated. Next, the translations are checked for bibliographic accuracy and completeness (missing figures, tables, references, etc.) As the Center has no professional translators on its staff, we make no attempts to check the accuracy of the translated text. The translations are then processed according to standard cataloging procedures. These may at times become quite involved, since a translation is a secondary product and the original foreign document may not be available for reference. Matters become complicated when the bibliographic reference to the original document is erroneous or incomplete. Each new journal title is verified in an authority source. When the bibliographic information is determined to be definitely incorrect, every effort is made to verify the correct citation. Fortunately we have the use of Crerar's extensive collection of foreign periodicals and monographs, enabling us to verify the translation against the original work. In cases when the actual foreign publication is not available, Crerar's unique collection of bibliographic and indexing tools makes it possible to establish the bibliographic entry and citation correctly identifying the original foreign document. After completion of this phase of the cataloging process, each translation is assigned a unique translation accession number and subject category according to the COSATI classification system. Standard catalog cards are produced at this point and the bibliographic data are stored in the computer memory for later retrieval for compilation of the monthly *TR-I* Index sections. The Index sections are cumulated semi-annually and annually. Also, catalog cards are filed in the appropriate master files. To faciliate searches for existing translations, master files are maintained by authors' names, journal citations, country of origin and patent number, report number, and translation accession number.

NTC offers a limited searching service. Upon receipt of the pertinent bibliographic information, the Center will search its master files to determine if and where a specific translation may be obtained. Journals translated cover-to-cover are included in the search. Searches for all translated works by a given author, or from a given journal are also available. Translation availability inquiries may be transmitted by mail, telephone (312-225-2526), or by TWX (910-221-5131). If available from the Center, copies of translations will be supplied at the cost of $15.00 for the first 1–10 pages, plus $3.00 for each additional 10 pages or fraction thereof. The searching service is operated on a fee basis. Organizations and individuals depositing translations are credited for the deposits by receiving free translation availability searches. Subscribers to *Translations Register-Index* are charged a discount rate search fee. Occasional users, who are neither depositors nor subscribers to *Translations Register-Index*, can obtain information searches at a fee of $5.00 per item.

The *Translations Register-Index* data base, accumulated since 1967, contains over 300,000 translation citations. In the past, the computer program has encountered serious difficulties caused by frequent turnover and loss of programmers and changes in hardware at the computer center where the work was done. Through the assistance of the staff and facilities of a major research center in the Chicago area, the program has now been reconstructed. To provide immediate access to information on existing translations, the Center proposes to publish a 15-year (1967–1981) cumulation to *Translations Register-Index*. The cumulative index would be a sequel to the *Consolidated Index of Translations into English (1953–1966)*. The two indexes jointly will cover all available translations since the beginning of listing of translations in 1953. The Center has the capability, knowledge, experience and dedication to produce the 15-year index. We sincerely believe that such an index will be of great value to the English speaking professional community. Due to its extremely limited budget, the Center is seeking financial assistance in realizing the project. Proceeds from the sale of the cumulative index will be used to further expand and develop the services and resources of the Center. Our final goal is to achieve online access to the translations information data base. Contracts have been made with several online services to determine the best approach. Publication and sale of the 15-year cumulative index will make online access a reality.

With the cooperation and support of the professional community served by the National Translations Center, the scope and area of activity is continuously expanding. Translations are our only access to the vast wealth of research around the world. They should not be neglected. Millions of dollars are spent annually by government research facilities and the private sector to translate scientific and technical papers. Unless the translations are made available to others and shared by all, we are wasting a wealth of information and an important resource.

The National Translations Center encourages all in need of translations to avail themselves of its services. Any questions or inquiries should be addressed directly to NTC.

THE ROLE OF COMMERCIAL TRANSLATION FIRMS IN PROVIDING TECHNICAL MATERIAL TO SCI-TECH LIBRARIES

Robert L. Draper

ABSTRACT. Increasing demand for translation of foreign literature has led sci-tech librarians to enlist professional translation services. Of these, the large, experienced bureau is likely to be more dependable and cost-efficient than in-house staffs or freelancers; yet even the most qualified agency may fall short of expectations. The process of translation can incorporate a number of quality/turnaround checks and balances, but any system works best when agency and librarian open effective lines of communication to both clarify and best meet the assigned tasks.

Over a decade ago, *Business Week* reported that a "healthy respect for foreign technological expertise" as well as the widening "international scope of corporate activities" had resulted in a booming demand among American businesses for translation services. Nothing has changed to minimize this demand; it would appear that the need to confront overseas competition as well as to "court foreign investors in their native tongues"[1] have only intensified.

Concurrently, trial-and-error (if nothing else) has established "professional" translation as an undeniable necessity. Attempts by amateurs to render "literal translations" of foreign documents lead to amusing or outright disastrous results;[2] the time and cost advantages, most requestors have discovered, are not worth the risks. "False cognates"—foreign words which might appear to resemble English terms but which have entirely different meanings—abound in all languages; supposedly comprehensive "polyglot" dictionaries may be of no help and often do more harm than good. Other "false friends" include poorly translated English summaries provided in foreign journal articles; scientific abstracts, which are usually reliable when used for their intended purpose (to aid information seekers in selecting relevant texts for perusal), obviously have inherent limitations in conveying

Robert L. Draper is Quality Control Director for the Ralph McElroy Company, Custom Translation Division, P.O. Box 7552, Austin, TX 78712.

to requestors all necessary information; and, for legal documents, *anyone* who is not an adept professional legal translator.

In seeking out professional translation services, a sci-tech librarian must consider the obvious needs—high quality, affordability and reliability—as well as circumstances which might create unusual needs. This article shall discuss all of the above, save price, since the cost of translation varies so enormously between agencies and fluctuates so drastically depending on the nature of the document that no useful range can be estimated.

The sci-tech librarian has three types of translation services from which to choose:

1. *Staff translators.* A handful of companies, including Du Pont and RCA, employ linguists and in-house translators, citing greater quality control and quicker service as prominent advantages to this approach.[3] Other realities, however, have turned most companies against the in-house system. The translator frequently is confronted with unrealistic demands and faces a continued need to ''educate'' management.[4] More tangibly, economic realities render the in-house notion unthinkable to companies which must address daily the inevitability of down-time and lay-offs. As the recession devastates R & D budgets, staff translators are sure to become economic casualties.

2. *Freelancers.* To avoid paying down-time, a librarian will sometimes employ the services of freelance translators on a job-by-job basis. The cost advantage gained by using outsiders with little or no overhead is outweighed by a number of disadvantages: the freelancer's capacity to handle a large volume and/or a variety of languages is generally limited; he/she typically cannot afford, as a freelancer, the proper amount of reference material required for quality translations;[5] deadline-responsiveness is hampered by the erratic flow of work; and he/she may farm out excess work to other freelancers, thus exacerbating problems of deadline and quality accountability. While a number of higher professional freelance translators do exist, none of them (at least as individuals) can truly meet a large company's complete multilingual high-volume needs.

3. *Bureaus.* Most companies rely on actual translation agencies to handle their workload, recognizing such bureaus to possess greater economic stability along with greater capabilities in terms of volume, language and subject matter, than freelancing individuals. In selecting such an agency, however, an information specialist must take note of the following:

—It only takes one office room and two individuals to constitute a "translation bureau"; indeed, many such agencies exist in a fly-by-night capacity, having no greater capability than that of a freelancer and with a questionable sense of professionalism. Truly professional agencies carry with them distinct reputations known throughout the information-gathering community; I strongly recommend that librarians rely on references from colleagues or that they thoroughly investigate the nature of the "agency" in question.

—Many agencies can't, or won't, respond to the urgency of a translation request. While the average turnaround time for translation of a standard document may vary from a week to a month or more between major agencies, requests for rush service are often met with outright arrogance. The vast majority of large agencies charge extra—as much as 25% or even 50%—for urgent attention; as the head of one of the country's largest agencies told me, "Yes, we do rush work, and yes, we charge the hell out of them for it." Such agencies complain that rush service undermines the quality of the translation, adding that the price mark-up for such service is always passed on to the translator as over-time compensation. Neither point is always true: most translation bureaus simply cannot adapt to strict deadlines, and usually because such agencies rely heavily on out-of-town freelancers rather than on an in-house staff. Librarians whose superiors demand immediacy, or at least punctuality, would do well to enlist the services of a translation bureau which can respond to such requests.

—Consistency in quality cannot be taken for granted. Even agencies who claim undying dedication to the quality of their work frequently have no control over their own product. In conducting telephone interviews with some of the nation's largest, most reputable translation firms, I found that while all respondents insisted that "none of our work goes unedited," conversations with their translators sometimes proved the opposite. Agencies for freelancers may rigorously test their applicants, but beyond this initial stage, such freelancers are told to provide "final-form" work which their agents simply slip into another envelope and send to their customers. Any translator, no matter how skilled, can make a mistake, and librarians are advised to select an agency which is set up for quality control and for the minimization of human error.

—Even a high-quality, low-cost, deadline-responsive translation bureau may not be enough for a librarian faced with peculiar needs.

Unusual assignments may require that a translation firm have literature-searching capabilities, printing facilities, audio-visual equipment, computers, mobile personnel and the like. Librarians may also seek an agency which is capable of translating confidential material, English-into-foreign material, enormous blocks of research and cover-to-cover foreign journals.

What becomes of a document between the time that a request for translation is issued and the time that it is returned to the requestor?—I consider it crucial that the librarian know. We must understand each other's needs, and they are great on both sides: both client and firm are confronted daily with unusual human and economic pressures, and it is just as naive for a translation bureau to assume that each customer expects exactly the same thing as it is for a customer to assume that the bureau can guess otherwise.

In discussing the translating process, it is necessary to point out that no one system exists: those used by the hundreds of translation firms in the U.S. differ from each other so sharply as to stagger the mind. Some firms employ an in-house staff of translators, while others do not; some utilize technical editors, while others rely on multilingual scientists; some purport to do any language and any subject, while others specialize; and the degree of technological advancement ranges from agencies with CAT (computer-assisted translation) schemes to those with only a couple of electric typewriters.[6]

I am obviously most familiar with the agency make-up of the Ralph McElroy Company, being employed as their Quality Control Director. Since the role of a translation firm in assisting the technical librarian depends on the firm's structure and attitudes, I shall occasionally allude to my company's system in comparison to those of other agencies. For now, a brief description of my agency might be useful—not for the purpose of establishing the Ralph McElroy Company as a definitive model, but rather to convey an image of bureau constitution.

The Ralph McElroy Company has one office; over 100 translators and editors are on its payroll, along with administrative and marketing personnel. About 85% of our work is done in-house, and this figure has been a constant over our 15-year history, despite enormous growth which has seen our 1981 gross business double the past year's figure. (We forecast a 1982 gross which will in turn double 1981's business.)

Our customer list mainly features those in the Fortune 500—large corporations with ongoing research concerns. We most frequently translate information involving chemistry, medicine, engineering, biology, geology, physics, etc., along with legal documents, confidential corporate and military reports, product specifications and advertising material. Nearly one-third of the orders we receive request translation of patents—61% of the patents being Japanese, which is our highest-volume language (20% of all orders) followed by German, Russian and French.[7]

In addition to our "custom translation" workload, the Ralph McElroy Company also provides cover-to-cover translations of *Khim. Prom.* (our title: *Soviet Chemical Industry*), *Melliand* (a German textile journal) and *Soviet Power Engineering*. Our English-into-foreign work has also tripled over the past year.

Many of the above-cited features typify the professional translation bureau of recent times: large, economically stable and interested in exploring new markets. For now, however, most companies of use to sci-tech librarians specialize in the custom translation work: new departments are built around it and new techniques maximize its "custom" development, but the custom division is the core of the translation agency.

An order for translation reaches a translation agency by mail, with or without prior arrangement. Our company and others similar to it have begun to expand into data-base literature searches, enabling us to procure the desired documents ourselves; otherwise, the literature is mailed with instructions specifying the nature of the work and the deadline for its return.

The process of assigning a translation, more than any other area, illustrates a particular agency's concept of responsibility, since the matching of document with translator reveals the agency's balance of concerns. A great deal has been written about the translator as a creature of solitude—"the lost children in the enchanted forest of literature," according to one manifesto[8] and the victims of more suicides and nervous breakdowns than any other European Community group of staffers[9]—but these images are as misleading as that evoked by the hapless U.N. translator in the television series "The Jeffersons."[10] A translator can be a bilinguist (or multilinguist) by accident of birth or by design; skilled technical translators, however, are a proud and intelligent lot, and all other generalizations are foolish.[11]

The skills necessary for a technical translator have also been discussed in never-ending fashion. Commentators seem to agree that such an individual must have a command of not only the foreign language in question, but of English as well; that the person should have all necessary resources for professional translation; and that he/she must have a technical background of the field in question.[12] The degree of the latter prerequisite has been debated *ad nauseum*: some agencies and individuals insist that only worthwhile technical translator is the so-called "multilingual scientist" (generally speaking, a technical expert by trade and a linguist by circumstances or secondary training),[13] while others tout the linguistic starting point after which technical expertise is gained on the job and with supervision.[14]

My experience has been that either type of translator can fulfill professional expectations, but only if they possess the following two interdependent qualities:

1. *A discernible will for knowledge and proficiency.* Technical translation is a skill forever influenced by the winds of change; unlike learning how to ride a bicycle, a translator of technical information must continually "re-learn" his/her field based on technological advances and terminological changes. Areas such as plastics, textiles and wood chemistry challenge the translator to be informed, to judiciously and creatively discover the re-defined present tense. No translation bureau can force this upon their employees—the will is theirs, or the absence of it breaks them over time.

2. *Honesty.* "To thine own self be true" should be the axiom for the technical translator: scores of "professionals" refuse to acknowledge their own limitations and are too proud to ask for help when the need (inevitably) arises. Unknown trade names, bizarre nomenclature, contextual ambiguities or areas which are simply "too technical" confront active translators almost daily. Whether or not they face their own shortcomings determines, better than any other yardstick, the good or bad translator. Our company, undoubtedly like others, has been willing to shoulder the cost of the honest and conscientious translator's struggles, but we have no patience for delusionaries.

From the file of such individuals a name is pulled—preferably one who has translated documents for the particular customer on the particular subject. The translator, equipped with necessary background material and standard technical references, begins his/her rendition, ideally communicating with the agency as to deadline adherence and any problems which might arise in the translation process. (I should mention that such problems are frequent and often are not the fault of the translator: the provided document photocopies are sometimes illegible, incomplete or replete with cryptic handwritten notations—all unfailing sources of confusion for the translator.)

Before, during or after the translating process, the decision regarding quality review is made. As mentioned earlier, some translation agencies do not even incorporate a mechanism for this possibility; the majority, however, at least grant the possibility (if not assume the inevitability) of circumstances which would require a technical editor to review the translator's work. A number of such circumstances exist: the document is unusually esoteric or unclear; a competent but "dishonest" translator cannot be trusted to admit his/her troubles with the document; or a less experienced translator, despite contentions that the translation was routine, simply may not know enough to be aware of problem areas. Our company has favored editing if for no other reason than the fact that "technical bilingual dictionaries," upon which many translators rely, are themselves frequently outdated, misleading or simply incorrect. [15]

Technical editing is performed by an individual who preferably has field (as opposed to merely academic) experience in the particular area; who has a "feel" for both linguistic nuance and technical writing; and who knows the translator and can thus not only communicate with him or her but can also anticipate problem areas. The intangibles which raise or raze translators apply equally to editors: he/she must have the creativity to discover and make use of all current data, as well as the humility to disclose to superiors unresolved problems and questionable passages.

Such problem areas may or may not be resolved, depending on agency structure, cost-efficiency and customer demands. The McElroy system allows for peculiarities, which are left to our quality control division to overcome. Examples include unconfirmed trade names which are not listed in standard references; passages in the text which appear to be technically incorrect; uncommon abbreviations; and documents which cite new technology and thus require consultation with the customer. The possibility of such consultation is a translator's ideal, but cannot always be utilized since agencies which rely on freelancers fear that the latter will steal the former's customers if company names are divulged.

After the translating and optional editing/quality-monitoring stages are completed, the rendition is typed, proofread, corrected, packaged and sent to the customer. Some agencies routinely follow up the product with a call to the customer to assess the reaction. Translation firms may also bill the customer at this time, although special projects and lengthy endeavors may require advance installments.

Translations are expensive to both the client and the labor-intensive translation agency, and it is to everyone's advantage that a communicative partnership be established. A librarian should feel welcome to question a translating approach, to request rush items or special format requirements, and to hold the agency accountable to whatever transgressions it might commit. At the same time, a translation firm should be able to request background material and consultation with the specific requestor, to ask for deadline extensions when obstacles prevail, and to explain linguistic ambiguities to the customer. This higher level of communication can only aid both parties, producing advantages such as the following:

1. The ability of a translation agency to minimize librarian errors by questioning the desirability of a particular translation—for instance, a request for translation of a document which, when reviewed, appears to bear no relevance to the client's field of interest;[16]
2. The ability of a translation agency to maximize quality by receiving word as to upcoming projects (which can then be met by anticipatory resource-gathering), stylistic or technical preferences, and positive feedback (which can institutionalize an agency's use of a particular translator for that customer);
3. The ability of a translation agency to review a customer's field of interest and to recommend a cover-to-cover translation of a related journal which the agency publishes at reduced cost per word; and
4. The ability of a translation agency to respond to deadlines by understanding real needs (e.g., an impending corporate summit) and to timetable purchase orders from clients.

The above examples serve to dismiss any notion that "a translation is a translation is a translation"—a notion, indeed, to which some information specialists adhere only until costly experience proves otherwise. Given the proper lines of communication, the professional translation bureau is eminently capable of bridging a client's conspicuous research gaps.

Almost every major American enterprise stands to benefit from understanding its global position via translation of foreign literature. Whether or not it will achieve that level of understanding depends not only on the selection of a truly professional translation agency, but also on its willingness to exploit the potential of the client-agency partnership.

REFERENCES

1. Up against the language barrier. *Business Week*. (2117): 164, 1970 March 28.

2. Affiliates of the American Translators Association frequently cite literal mistranslations in their newsletters. One chapter points out that the Russian literal interpretation of "pulling the wool over a foreigner's eyes" comes out to "hanging noodles from the visitor's ears"; another newsletter mentions a passage in a Polish travel brochure which an American might literally translate, "As for the tripe served you at the Hotel Monopol, you will be singing its praises to your grandchildren as you lie on your deathbed." The same newsletter points out "history's most tragic error in translation": In July 1945, the Japanese Prime Minister reacted to the Allied demands of unconditional surrender with a time-buying "no comment," or "mokusatsu" (derived from "silence"), which international press agencies misinterpreted as

"silent contempt." Irritated by the tone of the response, American leaders responded with severity and dropped the first atomic bomb ten days later. Editorial; *The Philadelphia Inquirer.* 10-A, 1980 August 2. Quoted by Olga Karkalas. *Delaware Valley Translators Association Newsletter.* 1–4; 1980 August-October. Noel, Claude. History's most tragic error in translation. *Association of Professional Translators Newsletter—Pittsburgh.* 5(4): 3–4; 1980.

3. *Business Week, op. cit.*, p. 164.

4. Tillinghast, Grace. In-house translators in industry. *The ATA Chronicle.* 10(10): 28–30; 1981 year-end.

5. Klein, Fred. U.S. trends in translation. *NOTA Bene.* 1(4): 5–6; 1979 Fall.

6. Klein, *op. cit.*, p. 5.

7. The prevalence of Japanese, German and Russian translations is typical of most agencies today. Twenty years ago, Special Libraries Association conducted a survey with regard to the most frequently requested technical information in languages other than English. The results placed German, French and Russian, respectively, as the top three languages, but also indicated that the demand for Japanese translations would increase while that for French would wane. SLA technical translation survey completed. *Publishers Weekly.* 181(16): 72; 1962 April 16.

8. Clements, Robert J. A bill of rights for wronged translators. *Saturday Review.* 53(25): 30; 1970 June 20.

9. Lewis, Peter. You like tomato and I like tomata. *Macleans.* 93(18): 14; 1980 September 29.

10. More than one ATA newsletter has lashed out at the demeaning portrayal of translators in this television series. See, for example: Bierman, Bernie. Letters to the editor. *The ATA Chronicle.* 10(6): 2; 1981 September.

11. Nor can translators be generalized by gender: between one-third and one-half of the nation's translators are women, and one expert names 10 women and 10 men as the nation's top translators. Richardson, Linda. Translating your skills into $35,000 a year. *Ms.* 7(17): 97–98+; 1979 November.

12. See for example: Gottfurcht, Adolph. The technical translator. *Chemistry.* 48(10): 15–16; 1975 October. Richardson, *op. cit.*, p. 97. Wright, Dr. Sue Ellen. Profile of a competent professional translator. *NOTA Bene.* 1–2; 1980.

13. Barb, W.G. The translation of German and French chemical texts into English. Source unknown, 566–572.

14. Richardson, *op. cit.*, p. 98.

15. Fred Klein, head of the German Department for Agnew Tech-Tran Inc. in California, points out that "Since a dictionary takes an average of 10 years' preparation, it suffers from built-in obsolescence." Klein, *op. cit.*, p. 5.

16. The problem of mistaken request is particularly acute in ordering Japanese patents, where examined (Koho) and unexamined (Kokai) documents employ the same numerical system for entirely different patents, meaning that a librarian may order a 1981 Koho document when a 1981 Kokai document was intended.

THE TRANSLATOR IN THE UNITED STATES

Fred Klein

ABSTRACT. Reviews the nature of the work of translators in the United States in recent years, including financial aspects, types of employment and commercial services available.

A presidential commission recently labeled America's incompetence in foreign languages as nothing short of scandalous. Many Europeans look down on Americans and criticize Uncle Sam as a bad linguist. Before rushing to uninformed judgments we should look at historical developments. Assurance of an unbiased viewpoint may be found in the fact that I was borne and educated in Central Europe where I lived for 26 years. After spending many years in Latin America I moved to California where I have resided since 1963, having traveled extensively.

The Melting Pot Syndrome

When the United States were established, the new nation adopted the language of the Colonies, English, as their own. A nation of immigrants needs *one language*, a unifying factor. Adventurers and refugees, pilgrims and pioneers landed at the shores of the New World: people of many lands wanted to forget their pitiful past, to merge into the mainstream. The immigrants spoke with a thick accent and their children, the second generation borne in the new country, were ashamed of them: moreover, in an attempt to appear more "American" they refused to learn the language of their forefathers. Only the third and fourth generations had lost those complexes and began to search for their ethnic past with pride. TV epics like "Roots" would have been unthinkable years ago. And until recently, before cheap air travel and global communications Americans lived in a certain geographic isolation. But theirs was a continent rather than a country, extending thousands of miles, with everybody speaking English. The industrial revolu-

Fred Klein is Head, German Department, and Chief Lexicographer at Agnew Tech-Tran Inc., P.O. Box 789, Woodland Hills, CA 91365. He has a BA from Officina Pragensis College in Prague.

This article is reprinted by permission from *Lebende Sprachen*, 1981, Vol. 26, No. 2 and the author.

tion made the U.S. the richest market in the world. But during both World Wars, young Americans came in contact with other cultures and languages. In the times of UNRRA and the Marshall Plan, a war ravaged Europe welcomed U.S. products and services uncritically. A world language emerged: English had become the language of international conferences, of scientific literature. It is the *only* language of international aviation. And American habits have extended world-wide, from hamburgers to Holiday Inns and Hilton Hotels. U.S. executives going to Germany find that almost all partners speak English—why should they learn German, after all? The illusion that English is spoken all over the world persists. Americans tend to view the French language oriented attempts of their Canadian neighbors with deep suspicion, as an anti-business, anti-American political movement.

Success Is Spelled English

The rapid rise of the U.S. to the status of a superpower with tremendous technological advances and a high living standard has brought about unexpected consequences. In the eyes of millions, America symbolizes success and power, regardless of recent failures. Somehow, on an *unconscious* level, this idea seems to translate into the notion that anything foreign is bad. A high State Department official said: In Anglo-American culture there is a bias against foreign languages and a bias against those who speak foreign languages fluently.—And South America?

Languages in Latin America

South American countries—likewise the product of immigrants speaking many languages—present a surprisingly different picture. Spaniards and Italians were the original settlers of what is now Argentina and Uruguay. Though the Spanish of the Colonies prevails as official language, Italian is spoken as well as German, English and French. Ethnic groups are proud of their heritage and do not hide it. The profession of the translator is in high esteem with "traductores públicos" to be found in all larger cities.

The United States-Lagging in Languages?

Europeans who grew up in a multilingual environment where many official languages coexist as equals in the European Community may view the situation in the United States as alarming, bewildering or puzzling at the very least. Europe's history of conquests and wars, invasions and occupations may also be interpreted as a history of mutual cultural and linguistic influences which extend across national borders. The United States on the other hand has fought during the last 100 years wars away from home but has not known foreign occupation in modern times. If foreigners point to the English only environment of the U.S. and comment on the lack of interaction with other cultures, Americans are quick to interject that the very fact of one language made possible the biggest market, the highest prosperity in history for a single nation, and their claim is justified. This background must be taken into account when analyzing translator problems.

The U.S. Marketplace

How much is a translator paid? How are the job opportunities? We must distinguish basic categories:

(a) International Organizations

The extraterritorial United Nations in New York have stringent requirements and very few translators or interpreters qualify for positions paying between $20,000 and $40,000 a year. There are few other organizations beyond the UN.

(b) U.S. Government

The Federal Government employs over 500 full time translators. A large volume of translation takes place through the Joint Publications Research Service (JPRS) which is the government's translation clearing house. Some professionals have called the JPRS a true mass production sweatshop. Translation rates were stated in 1979 as $15–40 per 1000 words and $4–12 per abstract. Under certain conditions, JPRS translators are paid another $5–7.50 per 1000 words for looking up specialized terms and/or reading up on specialized subjects, plus $4–6 per hour for processing tables, graphs and charts included in a translation. The Language Service Division of the State Department, employing some 90 full time translators, is also responsible for all translating requirements of the White House. Many other agencies are involved in translating activities, in particular the U.S. Air Force. The FBI, CIA, Department of Defense and other classified agencies use additional translators not included in the above figures. Salaries of staff translators working for the Federal Government can be at the starting level of GS7 (1979 salary range: About $14,000) and go to the top level GS13($38,000). Job titles are important: Designations as "information scientist" or "communication specialist" supposedly imply more than "translator."

(c) Private Sector In-house Translators

In-house translators may work either full time on translations only or they perform other activities as salaried employees and translate only when the need arises. Staffs—even in large corporations—tend to be small. Publishing houses do some limited in-house translating: the majority is subcontracted to freelancers. No established rates exist and remunerations vary widely.

(d) Freelancers

Freelance translators work either directly for their customer or for a translation agency on a part-time basis. These so-called independent contractors may charge rates starting as low as 3.5¢ going up to 35¢ per word. The American Association of Language Specialists suggests that translators be paid 4 cents per word as a starting fee. A known translation bureau pays freelance translators from $3 to $7.50 per 100 words, with higher rates being paid for translation from English into foreign languages. But rates fluctuate so much that these figures should be considered as non-representative samples.

(e) Translation Agencies

Translation agencies work with full time staffs of translators, editors, proofers and typesetters depending on the size of the company. Almost all use freelance translators or consultants on a part time basis. Only permanent employees have fringe benefits as social security, disability payments, unemployment compensation, paid holidays, sick leave, group insurance (health, accident, life) and paid vacations. Full time employees are also eligible for overtime if they are paid on an hourly basis. As freelancers do not have any of the above benefits, larger agencies tend to pay them somewhat higher rates to compensate for the lack of fringe benefits. Again, rates vary so much in this sector that they become almost meaningless. A small agency may never check or edit work performed by freelancers—there is the case of the so-called envelope-to-envelope agency: the unchecked translation is simply taken from the envelope, placed in another one and mailed to the customer. Such shabby, highly unprofessional practices do exist but must not be confused with the operations of larger, important agencies.

Large agencies perform all operations in-house and require even freelancers to work on the premises. Moreover, some agencies do not pay by word any more. A compensation of $20,000/year is considered a good salary for a full-time translator in an agency, and a first class professional who can translate at least two languages could command a salary of $25,000. Rates for free-lance translators working in a large agency start at $8/hour.

Comparisons: Proceed with Caution

It is dangerous or misleading to jump to conclusions based on remunerations in the U.S. and Germany. In a valid comparison the same type of job would have to be compared with percentages of expenses rather than based on the remuneration alone: a given percentage for food, housing, transportation, clothing, etc., leaving a certain amount for discretionary spending. My experience during interviews with professionals in comparable positions abroad shows that the U.S. living standard—with the exception of health care for persons under 65 and the risk of unemployment—appears to be much higher than in Germany. Most prices of daily necessities remain way under those in the Federal Republic. Though the buying power of the dollar has decreased dramatically abroad, it remains high in the U.S., even with present inflation rates. In other words, a translator in the same line of work may gain much more DM in Germany than his colleague in the States after converting the salary of the latter from $ into DM; nevertheless, the American lives cheaper and can afford more. Other dangerous conclusions deal with institutions. Institutions of the scope or sophistication of a Siemens Sprachendienst simply do not exist nor can a counterpart to the Bundessprachenamt be found in the U.S. The very fact that a large U.S. company has a language department must not be interpreted as something comparable to German organizations. There is no union for translators—though they might be members of one which includes them—nor do we have a cooperative of the InTra type here.

Professional Organizations

The bulk of U.S. translators comes from freelancing independent contractors which could be foreign college students and recent immigrants. Many newcomers see their language knowledge as the only marketable skill in the beginning. Most of them cannot be considered professionals and the turnover of these "temporary" translators is significant. There is no professional certification by the government to the degree which is common in Europe. The amount of foreign language schools within the whole of the giant United States is estimated at some 90, with some famous institutions indeed; nevertheless, a small number for a world power. The only professional organization representing translators on a national scale in the United States is the American Translators Association, or ATA for short. This U.S. affiliate of the F.I.T. was established in 1959 as a non-profit organization. Currently the ATA has some 1400 members which include famous linguists and professionals, both literary and technical translators. The ATA publishes a newsletter, organizes annual conventions and extends accreditation to translators who are active members. In addition, the organization works through different committees and makes a concerted effort to enhance the status of the translator which is a needed, worthwhile task. Other regional organizations exist, and those representing interpreters, but none of them of true national significance.

Again I must caution against premature conclusions. Statistically speaking, an organization of 1400 members in a nation of 220 million people is in fact the only organization representing translators. How representative is it indeed? It is estimated that 15,000 to 20,000 translators belong to some type of organization. And if you try to define professionals by advanced degrees or as graduates of language schools you run into problems in an environment which is paradoxical; On the one hand, the need for more education and degrees; on the other hand the pragmatic approach with puzzling features. I used to carry with me all my certificates and letters of recommendation during the early years. It took me a long time to learn that most Americans are pragmatists, to a degree which is unknown to most Europeans. An able applicant is offered a chance, degrees are not decisive and the most important thing—the system works! The self-made man is an American concept—and he carries no diploma, he may be a high school drop-out—but he is a success.

Interpreters

About six years ago, the Superior Court in Los Angeles was looking for court interpreters (Spanish) which were then offered $50 for a full day and $30 for half a day—the work could be in court or in prison and the interpreter would be "on call" without guaranteed full employment. The only prerequisites for applicants consisted of "six months of Spanish experience." Hundreds showed up for a written test which was followed much later by a second one for those who passed the first one. It is almost impossible to understand such a situation with a bare minimum of requirements from the standpoint of Europeans. However, according to new California laws (SB 420 and AB 2400) all interpreters working in the courts or for California state agencies will have to pass successfully state administered examinations, both written and oral. A California Court Interpreters and Translators Association (CCIA) has over 200 members and has been in existence for over five years.

On the federal level, the Federal Court Interpreters Act (Public Law 95–539) was signed by President Carter on October 28th, 1978. A certification test will be given by the Administrative Office of the U.S. Courts in Spanish, in 1980. Needless to say, the overall efforts to upgrade the status of interpreters continue to lag behind those in Europe. This situation results in a lack of incentives and upward mobility to a point. Things are changing.

Inflation and Imbalance in Exports

The very fact that a President's Commission on Foreign Languages was formed indicated the President's interest in the problem. Congressman Paul Simon spoke of the Helsinki Agreement which became the starting point for the President's Commission. Simon stated:

> We are the only country on the face of the earth where you can go 3000 miles and speak only one official language, and that in itself has caused some insularity. We have oceans on two sides and two friendly nations on two sides and that has added to insularity...only 8% of the colleges in the nation require foreign language for graduation...in the area of Russian, there are more teachers of English in the city of Moscow than there are students of Russian in the entire U.S....in the last 3 years, despite its increasing commercial importance, 102 American colleges and universities have dropped the teaching of German...somewhere between 1 and 1½% of...inflation was caused by the imbalance in exports and that, in part, has certainly been caused by our insensitivity to language and culture. You can buy anywhere, anytime, in your own language, you can't sell anywhere, anytime, in your language, and we have failed to do so.

The insularity Congressman Simon speaks of is very much in evidence. I had great trouble in convincing a large customer that there was no such thing as the "dollarmark" currency specified on his billing. Another one sent a document from Linz asking to translate into English what he believed to be in Australian (sic). Such problems are getting worse in advertising translations where the cultural barrier may be critical. Except for truly international advertising agencies with foreign subsidiaries, Madison Avenue copywriters continue to pour out clever, crisp ad copy full of the typical double-entendres which have made American advertising famous. In most cases, the translator is faced not only with an impossible deadline but with a complete lack of understanding: Typically American concepts defy translation, but the agency has already invested in full color separation and art which must not be changed. As professionals in the U.S. we can attempt to convince the customer that a problem exists and sometimes we win. On the other hand I recall a full page ad produced on Madison Avenue and translated locally into German for two top German and one Swiss magazine. In each case, the translator had not understood the clever double-entendre and produced a wooden, pointless headline instead. As a result, the German and Swiss representatives were up in arms and the U.S. executive in deep trouble for having entrusted the translation to Europeans who did not understand it.

How to Find Top Translators

Good translators are difficult to find in Europe: here it is almost impossible. To start with, I could look for technical translators who are ATA members. The 1979 edition of the Professional Services Directory lists a total of 511 ATA translators nationwide. Narrowing it down to the area of Greater Los Angeles with a population of some 8 million, and to my needs which are English to German, Technical & Scientific, I find only one name; the gentleman resides more than 60 miles from the place of work and even with our long commuting distances I could not expect him to drive so far. Translators like him rate themselves in the directory as very good or good, in other words, no test is involved. A certain percentage of ATA members are "accredited." These members have submitted to examinations by the ATA Accreditation Committee. Typically, translators are first given a practice Mini Test which is published in the ATA newsletter. The Mini Test I have seen is short and difficult. For the actual examination (which supposedly takes 3 hours) applicants may take any dictionaries they wish. Two persons who do not know the identity of the applicant evaluate the test. The names of members accredited by ATA are published in both the ATA newletter and in the Professional Services Directory.

At my place of work the practices vary. We look for translators through newspapers and other media. Applicants gather on a Saturday for a mini workshop at the translation agency; after a tour of the facilities followed by a slide presentation the work is explained to them with a subsequent question-and-answer period. Interested applicants are given a three-hour test under controlled conditions. First they must write a 300 word essay in their mother tongue dealing with translation problems. Next they have to translate 2 texts covering different technical subjects. Last they have a choice between 5 different subjects from which they are to select one which they know best. The key difference as to the ATA test is the fact that we hand to the applicants specific dictionaries—the same ones we would use ourselves—for each test section and insist strongly that applicants *must* use them. Theoretically they could bring their own—they never have—but we have found that we can best judge the attitudes of applicants by the degree they do or do not use the reference material. If somebody does not use the dictionaries during the test he will be a person with poor research habits. The tests are then evaluated by the head of the respective language department and a senior editor, and the personnel manager. Those who qualify are considered to meet our minimum requirements only and in many cases they do not last. Few recent college graduates meet our requirements: people from the Defense Language Institute, Presidio of Monterey, fare much better. Some of our best translators come from the Institute which has a predominantly foreign-born faculty.

Independent or In-house?

If the freelancer is a seasoned professional limiting himself to one language and a few subjects, if he can line up enough work he might do well with little or no overhead. He risks being undercut by those who work for ridiculous, low rates sometimes accepted by an unsuspecting customer who ignores that he is buying poor quality, buying trouble. But effective standards for quality do not exist. The in-house translator enjoys many advantages as an employee. Government or research-oriented employees in non-profit organizations often work without the stringent deadlines of the private sector and with better reference material. Quality control for these translations may be little or nonexistent. I would say that many people translate but there are few true translators to be found. The in-house translator has the advantage of specialization with direct in-house contact with engineers and technicians. Not all of them are experienced or trained. In an environment with little knowledge of language problems this can spell trouble. There is the person whom I shall call the baby native. He was born in Germany and his parents took him to the U.S. at the mature age of 6 months. And imagine, he even took a full month vacation in Germany last year. If this person has managed over these years to read German to a certain degree he might be considered an expert by many, a translator, simply because nobody else in the place knows German and people are naive. Why hire somebody else? We have Norbert, the native! This is not a cynical statement but rather a real situation. Another character is the European sales representative. He might be a top salesman and therefore *all* his opinions are accepted by management here. That does not mean that he knows how to write well in German, or that he is a translator. His credentials appear to be impeccable: A true native living in Germany, he knows his business. How can an American translator dare to fight with the ''expert'' for good German? Quite often management will let the marketing man abroad be the only judge of quality and the idea that this German may not write good style is inconceivable. In management thinking, being a native equals being a good translator. And it is easier to use a man on the payroll or the salesman instead of hiring an outsider or a specialized agency to do the job.

Quality Control and the Translation Agency

Quality in translation per se is an elusive subject and quality control even more so. Top agencies cannot survive without top quality for 2 reasons: Their main and often only business is translation. Moreover they have a large overhead and must charge higher prices. The best ones are true service organizations with word processing, complete art departments, foreign language typesetting, interpreting, audiovisual departments, to name a few. And translations is still at the core of all, it is THE PRODUCT—and must be of top quality. Bad translations don't improve if the typesetting is tremendous, if the full-color layout is just phantastic. Good translations are made by professionals who in turn need up-to-date reference material on the spot. There is never time for exhaustive research in university libraries and massive investment in reference material for the agency library is a must. In my place of work, Agnew Tech-Tran Incorporated, I have noticed a trend over many years: Five years ago, about 90% of our work was into English, mostly information transfer. Now the opposite is true. With growing inflation, our agency has indirectly assisted in the training of in-house translators. Companies who originally gave us work now try to make much of their translations in-house; often the quality is lower but the price is lower, too. This trend has not stopped our unbelievable growth, on the contrary—but we are now getting more and more of the most *difficult* material to translate and also short-notice jobs with impossible deadlines.

The cherished concept of specialization is almost impossible within the agency framework. We might have more repeat work in given areas as radar, but otherwise, our unpredictable, cyclical workload may involve subjects ranging from tanks to telephones, stock brokers to submarines, computers to copper mining, dental supplies to Disneyland. One thing only is always in short supply; time. In the words of one of our senior editors: We don't want it good, we want it *Thursday*. But in the final analysis, the success or failure of a translation into a foreign language is determined abroad, with critics scanning each word, and locally on the marketplace by satisfied repeat customers. Which procedures are used to safeguard a consistently good product?

Translation Procedures in a Large Agency

At Agnew Tech-Tran in Woodland Hills, California, we have an organization which may be small by European standards but is large for the U.S. Aside from marketing, accounting, production departments we do have separate language departments for German, French, Spanish, Russian, Chinese and Arabic. Each of these has its own reference library and there is a central reference library in addition; a proofing department, typesetting, word processing, a typing pool, audiovisual and interpreting facilities. The following is a breakdown of the mandatory steps for a translation into a foreign language: First the job is word-counted and estimated in marketing. The approved job is handled by the production department through all other steps. A control copy is made and retained in production. The actual job is described on a work order with estimated times and assigned to a native translator who prepares a rough translation which could be handwritten, typed or on mag cards. Next the rough is checked and corrected by the editor. The next step depends on the original—it could involve transcribing the edited text on mag cards from handwritten copy, or typesetting, or direct input in our word processing system. Regardless of the procedure in this step, the next one is proofing. Proofer A makes his corrections in red color and initials each page. Proofer B adds his corrections but uses blue color and initials likewise in blue. Next the double-proofed pages are given to the word processing department or to the typesetting facility respectively for corrections. Final proofing must be performed by a native speaker of the target language: he first checks off whether all corrections have been made—in yellow. Then he makes sure that no other errors—as wrong hyphenations—have been made. These are marked and corrected and the procedure continues until the job is ready—the final O.K. is his responsibility. We hope to have more and more translators working with mag cards and envision a future with the rough input on the screen through the keyboard, with the editor correcting on the screen, hopefully eliminating manual proofing phases—but this is still in the future.

Jobs involving more than 2 languages are handled by a coordinator. Large jobs are those which may involve different translators working in separate shifts. Our definitions are different—a job involving 5 full-time professionals is a large job. In such cases, a project manager is in charge. He may supervise a work-oriented glossary produced by word processing which is updated at short intervals—I call it a "living" glossary—and act as contact with the customer, for example. He would check actual work progress against estimates and decide discussions between editors, establish preferred technical terms etc.

Millions and Millions of Monolinguals

Which is the single problem which affects translators in the U.S. most? We live and work in an essentially monolingual marketplace. Every professional is affected by it, be it the freelancer, the employee or the agency translator. The impact varies: A top freelancer who types his translation and proofs it may be less handicapped than a large agency which operates on the basis of teamwork. How do you get enough typists, proofers, typesetters who know foreign languages in this environment? You find an American engineer who explains the technology in English but the customer has no bilingual expert at hand. It is possible to call up the German factory but the budget does not allow it. And there are limits to reference libraries, even on university level. As an example, I have a technical translation from Czech to English. As Czech is not in demand, I do not have good dictionaries and the large local university has only outdated one plus some poor old German-Czech books which are inadequate. Even with German and modern dictionaries, a specialized field or brand-new subject can become a problem. In Germany you find specialized magazines in foreign languages on newsstands even in small cities which can help you. In Los Angeles, in a large city, few newsstands carry foreign magazines at all and you will not find the type you need—there is no market. You can get online data from computer services, microfiches, any kind of information you want, but in English only. Short deadlines preclude gathering of information from abroad. You simply run against walls, you cannot invent terms, you have a deadline and a responsibility. I believe that top professionals from Europe would find it difficult if not impossible to adjust to such conditions. As for me, I see a constant challenge in what others may interpret as handicaps. I enjoy team-work in a changing environment. I am privileged to work with brilliant linguists, American linguists, in a stimulating environment of pioneers.

AUTHOR'S ADDENDUM*

The statements made in the paper "The Translator in the United States" are correct, but, as the data were collected in part in 1979, certain updates and corrections are in order. Also it should be kept in mind that the audience for the paper consisted chiefly of professional translators, linguists and teachers abroad, mainly in Germany, but also in France, Britain, Italy and Spain.

In 1981, fundamental changes took place at Agnew Tech-Tran, the large translation agency where the author of the article is employed. The existing word processing equipment was removed and replaced by a computerized system. All translations are now performed on the individual work stations, that is, on screens, with a CAT (Computer-Aided Translation) system used for work from English into French, German and Spanish. Thus the so-called mag cards have become a thing of the past, and other, far-reaching, changes have been implemented: the finished translation can be transmitted directly to the in-house typesetting operation, eliminating many steps, from re-keyboarding to multiple proofing. It is too early to judge this change completely—the conversion took place less than a year ago. The author has been engaged mainly in the terminology bank side in his position as Chief Lexicographer.

After three inflationary years, prices and salaries quoted before have changed. The U.S. government has developed, through the Office of Per-sonnel Management, a position-classification standard for Language Specialist Series GS-1040 (March 1980). Positions listed have a monthly salary range from Grade 5 ($855–1100) to Grade 15 ($4055 and up). Certification of interpreters by the government has made some progress.

The key issue not mentioned in the article (dealt with by the author extensively in other publications) is CAT, or Computer-Aided Translation. As librarians and other professionals dealing with high technology issues are very much aware of progress made in the field of EDP, and as they are usually users of online searches, a summary of the situation as seen from the standpoint of the translator, and user of translations, follows. In the past we have dealt with unrealistic expectations and sometimes bombastic, unproven statements by vendors of CAT SYSTEMS. What are the facts?

Briefly, CAT, or machine translation (MT) has been the subject of academic research in many countries since the early fifties. Abroad, the Russians were very active; in Germany, concurrent efforts of the government (Krollmann) and Brinkmann (Siemens) resulted in the creation of TEAM and LEXIS, the first workable multilingual terminology banks in the world. The European Community (Bachrach) began with a multilingual data bank at about the same time. The first large-scale CAT translation system in the U.S. was SYSTRAN by Dr. Toma, using, originally, a large mainframe. In this country, CAT efforts were started mainly by the Mormon Church and continued later separately by professionals who set up separate companies. SYSTRAN was, and is, used mainly for information-transfer-type massive translation from Russian to English, funded by the USAF and probably by other defense agencies. In the sixties, the government paid for a large investigation effort dealing with CAT, carried out by the National Science Foundation. The results proved to be devastating in the sense that official government support all but dried up. A single effort was made to set up a multilingual terminology bank at Carnegie Mellon University (TARGET); it never passed the experimental stage and finally failed. SYSTRAN is being used in Canada and has been acquired by the European Community; European universities have joined forces to generate EUROTRA, the European CAT-system of the eighties. The above is obviously not a complete listing of events, but rather a brief, incomplete report. Operational systems in the U.S. include SYSTRAN, WEIDNER and ALPS. Some known users of CAT are Xerox, Wang, NASA, USAF, to name a few at random.

What can be expected from CAT in its present stage? To start with, CAT systems—even the low priced ones—are out of reach of the freelance translator or small agency. Enhancements, which translate into price reductions, may come only from big multinational users, possibly from abroad, from Canada, or from the European Community. Medium range users, such as large companies, must evaluate carefully whether a CAT system will be cost-efficient at all and whether it will amortize itself over a large period of years. The major stumbling block is usually the quality of the unedited output. In many cases, both pre-editing and post-editing is necessary to arrive at a product with acceptable quality level. This quality—AQL for short— has become the topic of endless discussions. Many promoters of CAT products claim that the readability is sufficient and that an *understandable* text is sufficient in most cases. What they do not always take into account is the extreme complexity of some text. This in turn necessitates often pre-editing into the so-called controlled English, that is, time and money, *and* extensive and expensive post-editing efforts. Unless the USAF SYSTRAN information transfer quality is considered sufficient, the product is too expensive if one uses CAT in its present form. If a raw output for information purposes is needed and massive amount of text processing is mandatory, then SYSTRAN or the like is ideal.

For a company to have a cost-efficient CAT system, certain preconditions must be met. These are:

  a. A relatively reduced "corpus" in a sublanguage.
  b. A significant ratio of high frequency terms.
  c. A relatively simple text with many repeating phrases, void of idioms.
  d. A large amount of similar text to be translated and retranslated over many years with minimum variations and/or additions.
  e. The existence of an online multilingual glossary or access to a terminology bank.

As to (a), a sublanguage is defined as a subject-specific group of terms for a given subject and in a given context. The best example is the weather forecast issued in English in Canada, and broadcasted 20 minutes later all over the country in French. This, probably the only semi-automatic translation in the world, still requires some intervention (about 20%). It is important to realize the relevant features of weather forecasting, namely the relatively limited, simple vocabulary which is *predictable*, which can be extracted by computerized methods from past forecasts, and can be assembled into a group of terms and idioms which practically repeat themselves endlessly in many variations. Here the human is relieved from tedious, repetitive tasks indeed, and the translation will pay for itself. However, the CAT system called TAUM, developed by the University of Montreal, was subsequently applied to a larger and much more complex task, the translation of aircraft manuals. As of now, the researchers were unable to cope with the increased scope and have not produced practical results.

The last—and maybe most important—problem of CAT involves online glossaries or dictionaries. It is usually dealt with very quickly in the literature of the promoters; they say correctly that the system will list missing terms which have to be input into the CAT system. Figures are important: For a CAT system, *each and every word*—even if it appears only once in the source text—must be input, coded and accepted by the system. That it must be *researched* before by a translator, is often omitted. We deal with a very costly operation. General Electric of Canada gives a value of $32 per word pair, an amount which will naturally vary up and down depending on the quality requirements. But in the U.S., we do not have a *single* large terminology bank; private ones serve the limited purposes of certain companies and seldom offer the material. Thus it is needed to assemble the sublanguage terms—all of them—to enter them *plus* each common word occurring. The ratio is crucial: If a 30,000 word translation contains some 13,000 words to be entered, the operation becomes inefficient as to cost and time, unless a guarantee exists that the initial investment in time and money—the words— will be amortized in subsequent massive translations of, say 300,000 words plus it will contain essentially the same words. Contrary to the human, who makes use of his/her memory, and thus is able to supply both words and syntax almost instantly, the computer depends on the fact whether the programmer has been able to put into rules the nature of language and translation *and* whether he has been given every word. Any CAT system in existence today still depends on every word or idiom entered and is "judging" simply words or idioms. It judges a sentence as a unit but is unable to ascertain relationships or meanings without human intervention. The latter is feasible in interactive mode but quite costly.

CUTTA SERVES:
ORGANIZING TRANSLATION WITHIN A UNIVERSITY

Victor Hertz

ABSTRACT. Describes a self-supporting activity at Columbia University which provides tutoring and translation service to persons in the metropolitan area. Sci-tech subjects play a large role in translations. Names of languages handled, quality control of service and financial aspects are discussed.

CUTTA is presently in its fourth year of successful operation. What is CUTTA? It is the Columbia University Tutoring and Translating Agency, a self-supporting division of the Office of Student Employment. The agency was established in 1978 with two purposes in mind: to provide income for university students and personnel; and to provide quality tutoring and translating at reasonable rates to persons throughout the metropolitan area. By registering with the agency, students can earn money while working in areas related to their academic studies.

The following discussion of the origins and operation of CUTTA illustrates a unique approach to the problems and possibilities of organized, professional translation. In general, working within the context of the university has been both a boon and a burden. Specifically, I believe that the story of CUTTA has relevance not only for those interested in duplicating this experiment in other universities, but also for professional translators (and their clients) who must contend with the dual nature of translation as an art or service and as a business. In any case, a good place to begin is an examination of what CUTTA does (or doesn't) do.

The Present Structure and Function of the Agency

CUTTA must, first and foremost, survive, coping with an environment which, at times, is difficult, if not hostile. As presently structured, the agency is a self-supporting entity. In this regard it resembles a business. At the same

Mr. Hertz is Manager, Tutoring and Translating Agency, Lewisohn Hall, Columbia University, New York, NY 10027. He has received from Columbia University the BA, MA, and MPhil degrees and is currently working on a doctorate there.

time, however, the agency is constrained to operate on a cash-flow basis and must annually balance its accounts without recourse to long-term borrowing. All expenses, from paper clips to typewriters, from advertising to staff salaries, must be met by charging fees for the services offered. Survival therefore means generating income on a more-or-less steady basis by satisfying the needs of a diverse clientele.

CUTTA's clientele includes three overlapping groups: the agency staff; the university community; and the metropolitan community. Each of these groups has very different needs, but it is the synergy of these different needs which allows the agency to function as well as it does.

For the agency staff, CUTTA is primarily a source of employment. Students who are registered as translators (or tutors) gain assured job opportunities, and a fixed rate of pay at a relatively high level. For example, tutors usually receive between $8.00 and $20.00 per hour; interpreters usually receive between $12.00 and $30.00 per hour. The agency also provides other support services including foreign language typing, centralized billing, and professional advice. Workshops are periodically organized to help train translators (and tutors), and translations are, as a matter of course, critiqued and edited. Both the workshops and the critiques are a valuable source of feedback for the budding professional. Indeed, the agency has the major attraction of creating jobs which parallel and enhance areas of academic study.

For the university community, CUTTA is a convenient source of quality translation at reasonable prices, including the translation and certification of documents. As measured in terms of income generated, however, the biggest need of the university community is in the area of tutoring. For students who want help in their language (or non-language) studies, the agency screens and trains tutors both for long-term assistance and "crisis" pre-examination "reassurance." For some students the latter is especially vital—helping them to cope with the demands of a high-pressure academic environment.

These same services, quality translation and tutoring, are offered to the metropolitan community at large. As in the university community, the primary concerns of the clientele are speed, reliability, and price. The emphasis, however, is much more evenly divided between tutoring and translating. In translation, CUTTA specializes in technical work, especially in the fields of engineering, medicine, and finance. Translations in these areas represent the bulk (approximately 65%) of our business from the metropolitan community. Among the list of our prestigious clients are members of the Fortune 500, leading research corporations, and international law firms.

The survival of CUTTA demands that we satisfy each of these significant clientele. In doing so, the agency assures not only its immediate source
of income, but also generates the good will needed for repeat business.
Without this good will the agency would be unable to continue, with it the
agency prospers. Moreover, it has the added benefit of generating good will
for the university as a whole. This latter benefit is invaluable in serving a
"hidden" fourth clientele of the agency—the university administration.

Origins and Growth of the Agency

CUTTA was founded in the fall of 1978 when I was approached by the
newly appointed Assistant Director of the Office of Student Employment,
Mr. Eugene Kisluk, with a suggestion that I organize a service under the
auspices of his office. As long-standing friends, I had broached with him
on a previous occasion the idea of starting a tutoring agency. Now, from
the perspective of his new position, he thought such an enterprise would
provide a low-cost mechanism for supplying valuable services under a theory
of student self-help. His major contribution to the effort at this stage was
to prod me into expanding the concept for a tutoring agency to include
translating as well.

I agreed to explore the idea, and approached a fellow graduate student,
Mr. Martin Newhouse, who is an accomplished German linguist and
historian, to join me in the venture. The attraction of what we all viewed
as an experiment was threefold. First it was to generate income through a
part-time commitment of time that would allow Mr. Newhouse and myself
to continue our academic pursuits. Second, it was not only a "business,"
but also a service that would help both students acting as tutors, and
translators and their clients. Third, it was an "adventure" in organizing and
applying scholarship, both teaching and translating, to concrete needs outside a classroom.

From the outset, CUTTA was something of an anomaly. While the two
managers were given *carte blanche* to initiate and operate what was, in fact,
a small business, neither they, nor the university, were entirely comfortable
with this responsibility. Although both Mr. Kisluk and his immediate
superior, Mr. Robert Gallione, were supportive, there was a general lack
of resources—especially a lack of seed money, staff, and space. As a result,
during the course of its first year, the agency focused on its tutoring, rather
than its translating, business; a tutoring agency simply did not necessitate
the high level of commitment of time and money needed to perform quality
translation. Only gradually did the success of the tutoring side of the agency

generate sufficient income to allow a slow but steady expansion. But as the agency grew, it became clear that the experiment would require a full-time (or more than full-time) commitment from the managers and a larger commitment (minimally in terms of office space) from the university.

In the summer of 1979, I took on sole responsibility for managing the agency. (Mr. Newhouse went on to complete his dissertation and obtain a teaching position in Massachusetts.) During the academic year 1979–80, CUTTA was able to move its operations into an expanded facility. At this time the agency also launched a major advertising campaign, evolved new procedures for the training and screening of tutors and translators, and increased its office staffing. The resulting growth during that year and in subsequent years has been gratifying.

Since its inception, the agency has provided more than 1,500 students with jobs; its aggregated earnings for registered tutors and translators are in excess of $500,000.00. It is, in fact the most rapidly growing source of student aid on campus. This is particularly important in view of the impending cutbacks in the federal education budget. The following table illustrates CUTTA's strong pattern of growth.

EARNINGS OF TUTORS AND TRANSLATORS (by Academic Year)

1978–79	1979–80	1980–81	1981–82
$62,000	$130,000	$171,000	$280,000*

*Estimated. Earnings from September to June were $238.000.

Still, the fundamental contradiction between operating as a small business and operating as a student service remains unresolved. Without access to long-term borrowing, which would be normal for a business, or to a direct university subsidy, which would be normal for a service, the growth of CUTTA has undoubtedly been slowed. Moreover, as the agency grows in spite of these shortcomings, the contradiction between business and service becomes an increasingly serious drag on the agency's continued success, most especially damaging the highly demanding areas of quality translation.

Technical Translations

Of particular interest in the context of this special issue of *Science & Technology Libraries* is our translation work in the technical fields of engineering, medicine, and finance. Although some professional translators denigrate technical, as opposed to literary, translations, it is the mainstay of CUTTA's business, and we have therefore devoted considerable effort to developing these technical capabilities.

Our niche in the New York translation marketplace depends on four separate factors. First is our pricing policy. For the past four years, I have tried to minimize the cost of our services. As a rule of thumb, the agency charges one-half to two-thirds the price of other established agencies in the metropolitan area. In fact, I often encourage clients to "shop around" and compare prices (e.g., with Berlitz or Interlingua) so that they can appreciate the discount we offer. The latter has been useful in educating clients to the standard price of quality translation.

Unfortunately, our low prices have raised the unexpected problem of confirming the prejudice of some of our clients that there must be something wrong with the CUTTA product: after all, underpaid and therefore inexperienced "students" must be doing the work. This prejudice survives, in spite of the fact that the same "students" who translate for CUTTA also translate for other established agencies. The real savings to our clientele comes from CUTTA's minimal mark-up. While most established agencies use a 100% (or greater) mark-up in setting prices, CUTTA has a mark-up of 10% to 50%. Cost-savings are made possible by our centralized, low-rent location near our source of translators, and our heavy reliance on word-of-mouth advertising. While our translators often receive more money working for CUTTA than they do working for other agencies, our low mark-up still allows us to beat the competition's prices.

CUTTA's second advantage in the market place is the range of languages in which we offer service. Because of the nature of the university, we have available native speakers of most major languages, many of whom are taking advanced degrees in economics, business, engineering or law. Moreover, the university has one of the most wide-ranging offerings of language courses anywhere in the world. For more esoteric languages, CUTTA is the only organized source of translators in the city. Listed below is just a sample of the languages in which services are currently available:

Translation Available in These and Other Languages:

African Languages	Latin
Albanian	Marathi
Amharic	Malay
Arabic	Norse
Armenian	Norwegian
Bengali	Polish
Chinese	Portuguese

Creole	Punjab
Danish	Pushto
Dutch	Romanian
Farsi (Persian/Iranian)	Russian
French	Serbo-Croatian
German	Spanish
Greek (Ancient/Modern)	Swahili
Gujarati	Swedish
Hawaiian	Tagalog
Hebrew	Thai
Hindi	Turkish
Hungarian	Ukrainian
Ibanag	Urdu
Indonesian	Vietnamese
Italian	Welsh
Japanese	Yiddish
Korean	Yugoslavian

CUTTA's third advantage is the speed with which we will turn around translations. In part, because our clientele is still numerically small, in part because "qualified" translators (especially in the romance languages, German and Japanese) are under-employed, CUTTA offers a "rush" service wherein a client may, for a small premium, obtain same day or 48-hour service on small manuscripts (3,000 words or less). Rush service is also available for larger manuscripts, but with comparatively longer deadlines. Moreover, it is policy to try to accommodate clients even at the expense of late nights and short week-ends.

CUTTA's fourth advantage is the reputation of Columbia University in the City of New York. This reputation often wins us the opportunity for a "trial run" with new clients. Referrals from university alumni and faculty also play a role in steering clients to the agency. Finally, the agency has now begun to develop a reputation in its own right as a "place to go" for translation and has drawn new clients through word-of-mouth.

All of these advantages do not, however, substitute for the *sine qua non* of translation quality. We at CUTTA are especially sensitive to problems of quality control, not only because the agency relies so heavily on word-of-mouth advertising and repeat requests for business, but also because of the prejudice against "student" translators. To insure quality the following three policies have been adopted. First, all translators must demonstrate their capability before being assigned to a job. This demonstration is either through past performance with the agency and/or through producing sample translations—preferably using text from the job at hand. Second, translators should, whenever possible, be "native speakers" of the target language. Although there are exceptions, it has been my experience that few people are truly "bilingual." The idioms and nuances of a language simply are not learned through a textbook, or even through travel, but require extensive and repeated exposure to educated native speakers. This seems to require years of living in the country where a language is spoken in order to develop the proper "ear." Third, whenever a relatively inexperienced individual is assigned to a job, he or she is teamed with one or more highly experienced translators. The experienced translator will sometimes be a faculty member, sometimes an alumnus, and sometimes, when it is warranted, a professional translator from outside the university community.

Quality control also extends to the process by which CUTTA translations are performed. The major focus of our effort is the formation of a job team. For a technical translation this may include: a translator, an editor, a technical editor, a typist, and a copy editor (proofreader). Sometimes two or three of these functions (e.g., technical editor and editor, or translator and typist) will be assumed by the same individual, but we always seek to assign at least three individuals to any major job. At the same time, whenever we are not under severe time pressure, we avoid assigning more than one translator, or more than one editor, since this results in an inconsistent style. In all cases the editor is the team leader and is usually the most experienced translator on the job. He or she is responsible for the flow of work and insures that deadlines are met. These and other responsibilities are defined in our standard guide sheets, excerpts from which appear below:

TRANSLATOR'S RESPONSIBILITIES: The translator should produce the best possible draft as per the specifications of the job. The draft must be *legible, accurate,* and *grammatically correct*. In all cases the translation should read well in the target language. The translator must redo, in a timely fashion, any portion of the translation which is rejected or risk nonpayment for the job.

FORMATING: The translator must submit copy which follows the format of the original in such a way that a typist, unfamiliar with the text, will have no difficulty in spotting changes in style (e.g., capitalization, indentation, punctuation). Paragraphs which are repeated should be photocopied and attached in the appropriate location in the text. Pages of the translation should be stapled to the corresponding copied page of the original.

EDITOR'S RESPONSIBILITIES: The editor is responsible for the quality of the final product. Corrections must be readable and legible, and in all cases formating guidelines which apply to the translator apply to the editor as well. The original text (or photocopy of the original) should be checked against the translator's draft to guard against omissions and errors in pagination, numbering and punctuation. The editor should work directly with the translator to insure that job specifications are achieved. Additionally, he/she has the right and the responsibility to rewrite portions of the translation which he/she judges awkward, inaccurate, ambiguous, or ungrammatical. Where severe problems exist, the editor should consult with the Agency.

TECHNICAL EDITOR'S RESPONSIBILITIES: The technical editor is responsible for the quality of the technical language of the final product. The choice of technical language (legal, scientific, commercial, etc.) must be appropriate, consistent and accurate. The technical editor should meet with the translator and/or editor prior to the completion of the first draft to insure correct usage of technical terms. He/she should edit the final draft before submission to the typist. On occasion, assistance may be necessary in the preparation of a glossary of technical terms.

COPY EDITOR'S RESPONSIBILITIES: The copy editor is responsible for proof-reading the final copy and making corrections in spelling, grammar, inconsistencies of language, punctuation, and syllabification. He/she should check the final copy against the original text to insure that the format is correct and that there are no omissions. The copy editor should work with the typist to guarantee that corrections are made properly.

In most instances of technical translation, the translator and editor are instructed to use a "literal style" in which accuracy and precision of language are the main objectives. The translation should follow the original as closely as possible without sacrificing readability. We distinguish this from a "literary translation" in which it is essential that the finished translation reads smoothly. When a "literary translation" is used, the main emphasis is on avoiding awkward language. The final product must follow the accepted style of the target language (e.g., Italian and Spanish allow for greater circumlocution than does English). Liberties may be taken with the text for reasons of style, but the sense of the original must be maintained.

Of course, it is something of a grey area as to when a text becomes literary rather than literal. No hard and fast rule exists, but in dealing with technical material the agency advises translators to emphasize precision rather than style. This is undoubtedly the most subjective aspect of any translation and ultimately it is the responsibility of the staff to carefully uphold the client's wishes and accommodate a client's needs.

Prospects and Possibilities

It is feasible that during the next academic year the gross receipts of CUTTA will increase to $550,000. Based on the record of the past six months, we project gross income generated by the translating side of the agency will grow to $300,000. Clearly, the volume of business for translation now outweighs the business generated for tutoring. A consequence of this development is an internal subsidy for the tutoring side of the agency. This reverses the original situation in which tutoring subsidized and allows the expansion of the translation side of the agency. Thus both the agency and the university have benefited and continue to benefit from the synergy of the combined tutoring and translating functions. The resulting growth in volume may, in turn, require further expansion of CUTTA facilities (e.g., office space, staff and mechanical facilities such as a word-processor). Whether or not such expansion takes place depends on how, or even if, the longstanding contradiction between CUTTA as a business and CUTTA as a service is resolved.

Can the organization of a tutoring and translating agency be duplicated at other universities? There seems to me to be three conditions which help insure the success of such a venture. First, the university should be large enough to support strong language departments, especially in French, Spanish, Italian, German, and Japanese. It is also desirable to have a wide range of other languages available. Second, the university should be located in, or near, a metropolitan area where there exists an established market for translation services. Third, the host university should, at the outset of the enterprise, make a firm commitment to the success of the venture: including space, staffing and sufficient funds for a sustained advertising campaign.

The CUTTA experiment can, to this date, be judged a success. The Agency has, in particular, benefited from the enormous academic resources of Columbia and from access to the market place of New York City. But I firmly believe that even a medium-sized university, in a relatively small community, could sustain a tutoring agency with some tie-in to translation. The potential benefits of such an enterprise are apparent. CUTTA is an experiment which welcomes replication.

EI'S INSIDE LOOK AT TECHNICAL TRANSLATION

Zoran Nedic
Barbara S. McCoy

ABSTRACT. Aspects of translating and abstracting technical literature are surveyed, using specific examples to show how a New York-based information service, Engineering Information, Inc. (Ei), accomplishes these tasks. Special emphasis is placed on the requirements and practices of technical translation in various engineering fields covered by Ei's services.

Introduction

While the intellectual process involved in translating remains rather ill-defined and receives only scant recognition, there exists little doubt as to its final outcome: a clear and effective transfer of ideas and information from one language to another. Even though this goal would be generally applicable to any type and form of translation, it assumes added requirements and tasks in the case of technical translation, particularly in the many specialized disciplines of engineering.

In this article, the authors will make a concerted effort to pinpoint and explain what these tasks are, and how they are accomplished at Engineering Information, Inc. Ei, a not-for-profit information service, has long-standing experience (since 1884, under its former name of Engineering Index, Inc.) in providing technical information to the engineering community throughout the world.

Composition of the Ei Database

Before delving into the more specific aspects of technical translation at Ei, an overview is in order of this institution's comprehensive databases, and the role played by the coverage of non-English language publications.

Ei, as an abstracting and indexing service, monitors over 4,000 primary

Zoran Nedic is Editorial Supervisor at Engineering Information, Inc., 345 E. 47th St., New York, NY 10017. Barbara S. McCoy is Communication Services Manager at Ei.

sources from around the world. At present, publications from over 40 different countries are reviewed. Forty-five percent of the primary sources are published in the U.S., 16% in the United Kingdom, 10% in the Federal Republic of Germany, 5% each in Japan, France, and the Netherlands, 3% each in the U.S.S.R. and Canada, with the remaining 8% from various countries around the world. The indexed sources are published in over 20 languages including 8% German, 4% French, 5% Russian, and 11% in various other languages. Approximately 72% of the indexed publications are in English with this amount including cover-to-cover published translations of non-English language documents.

In 1982, abstracts of approximately 115,000 individual articles will be added to *The Engineering Index* and COMPENDEX, the corresponding online database. These abstracts are drawn from journals; transactions, reports and special publications of engineering societies, scientific and technical associations, government agencies, universities, research institutions and industrial organizations; monographs; standards; and reports. An additional 100,000 articles from over 2000 conferences will be indexed in the *Ei Engineering Meetings* database, at present available only in a machine-readable format and online through vendor systems.

As of June 1982, the Ei database consisted of bibliographic citations, abstracts and indexing of approximately 55% journal articles, 35% conference proceedings (both reviews and individual articles), and 10% reports, standards and monographs. Since July 1982, coverage of individual papers from worldwide published conference proceedings has been increased and concentrated in the *Ei Engineering Meetings* database. As a result of this concentration, the printed publications, *The Engineering Index Monthly, The Engineering Index Annual* and COMPENDEX, the corresponding online database, will consist largely of articles from worldwide journal literature.

Whenever a non-English article is indexed and abstracted, Ei includes the original language title, and a translated English language title as well as an indication of the original language. The original language of a document appears as the last element in the abstract in COMPENDEX and *The Engineering Index*, and *not* as a separate element. In the *Ei Engineering Meetings* database, there is a discrete language field. If the original language of a publication is in a Cyrillic or Greek alphabet, Ei transliterates *and* translates the title. If the original language is a Oriental alphabet such as Chinese or Japanese, Ei provides only an English translation of the title. Neither the original nor a transliterated title is provided. In cases where only an English language title appears, Ei gives the English language title as the original language title.

The retrieval of articles by language in COMPENDEX is influenced by these policies. Of the 8 online vendor systems on which COMPENDEX is presently available, 4 offer language as a search element. However Ei will be introducing a language field to COMPENDEX in 1983 which will allow language to be searchable on all vendor systems.

Retrieval of non-English language documents by searching the new *Ei Engineering Meetings* database is straightforward since the language of the original article is in a discrete field and is directly searchable on all available vendor systems.

Format, Style and Editorial Content of Translations

Ei provides a concise, factual and easy-to-read English-language abstract of selected primary documents. Editorial content of this English-language abstract will faithfully represent the viewpoint of the author(s) of the original document without added information or quality judgement. Ei will abstract only those documents that are of foremost interest to the engineering community, and that offer excellent quality, timeliness, lasting value and are of a high technical level. When translating the title of an article into English, Ei editors strive to stay as close as possible to the original title without translating literally or sacrificing the clarity of meaning of the title. Both the translated titles and text of the abstract are rendered in standard acceptable technical English.

Editorial policies are thoroughly implemented at every stage of translation work. Above all, this is reflected in the editor's decisions about the article's main emphasis, the inclusion of relevant technical facts and figures, as well as the choice of most appropriate indexing terms. When covering a non-English journal editors assume primary responsibility for the selection of articles for coverage.

A Profile of Ei Translators/Editors

All translating and abstracting services involving non-English language publication at Ei are the responsibility of its editors. The background of the Ei editors is quite diverse: In most cases, it embraces both technical and linguistic capabilities. Most Ei editors have engineering degrees and, at the same time, are proficient in several foreign languages. A number of Ei editors have an academic background in languages and are also well-versed in editing technical publications dealing with specific engineering disciplines. The number of languages in which Ei editors are proficient is large and continues to grow. It includes the most widely used non-English languages, French, German, and Russian, but also, Italian, Dutch, Portuguese, Spanish, Ukrainian, Finnish, Estonian, Greek, Hebrew, Romanian, Chinese and Japanese, among others. The Ei editors are familiar with Latin and classic Greek which is especially useful when dealing with special symbols and phrases encountered in technical literature.

The editors that handle non-English publications have a dual and complementary role. They perform selection and indexing duties which are assigned to all publication editors and will also translate non-English documents. This role enables them to keep up-to-date with the latest terminology used in English-language publications covering the most recent engineering developments and technological breakthroughs. This valuable knowledge is then applied to their translating and abstracting work on non-English language documents, allowing them to use the most current terminology in abstracting and indexing

Mechanics of Technical Translation at Ei

The main groups of publications that are covered by Ei are: (A) serial publications, including professional and trade journals, and (B) non-serial publications such as standards, papers, monographs, and special reports. Each editor is assigned a number of serial publications to review on a regular basis and also receives various non-serial publications as assigned by the editorial supervisors. Assignments are made on the basis of an editor's area of technical competence, the editor's language capabilities, and Ei's requirements relating to its subject and language coverage.

Selection of Non-English Language Documents

Once an editor receives a non-English language publication, several tasks are performed before the actual translation takes place and the abstract is prepared. First of all, the editor selects which papers contained in the publication will be indexed in the Ei databases; i.e., whether a complete or selected coverage of the publication is warranted. In some instances, Ei provides cover-to-cover processing of all technical articles from journals and conferences. In the case of those journals which an editor receives on a regular basis but which are not processed on a cover-to-cover basis, the usual Ei criteria of selection, mentioned earlier, are used as guidelines. Editors are often familiar with a particular publication, and know the types of articles that will appear as far as the technical quality, editorial content and relevancy are concerned. Therefore a judicious decision as to coverage can be made by an editor rather quickly.

Focus on the Translating Process

Next lies the task of actually translating the text and preparing an English abstract of the article that has been selected for coverage. To do this, the editor scans the article to determine its main thrust and key points, takes notice of any references and/or bibliography, and translates the title. The technical complexity of the subject requires the editor to carefully read the text in its entirety in order to comprehend its full scope. This is especially true in the case of articles that cover new processes and products, as well as recent engineering and technological discoveries.

Finally, the editor creates an English abstract, accurately representing the content of the original document. At times, the original document is already accompanied by a printed abstract in English. In that case, if the provided abstract faithfully represents the original article, is written in acceptable English and otherwise meets the Ei editorial criteria, the editor may adapt the original abstract. The English language abstract which accompanies a non-English article may be written in grammatically incorrect English, or may not be intelligible, or can occasionally misconstrue the author's point of view and therefore may not be adapted. If the editor finds the abstract unsuitable, he prepares another abstract that conforms to the Ei standards.

When translating, abstracting and indexing a non-English language article, the editors utilize the most current technical and scientific dictionaries, reference books, manuals, glossaries, and encyclopedias to determine not only the meaning of a term but also to determine the most widely accepted translation of a term or phrase. Some Ei editors also compile their own specialized glossaries of technical terms, acronyms and abbreviations. The editors also have easy access to the entire collection of the Engineering Societies Library (ESL) that is also located in the United Engineering Center in New York.

Perhaps the most valuable aid available to Ei editors is the consulting process with each other to clarify a technical term or to gain full understanding of a new technical process or a novel concept appearing for the first time. There are a number of Ei editors who cover literature of the same or overlapping engineering disciplines, and they frequently consult each other on such matters. For example, if a metallurgy editor is covering an article dealing with physical and mechanical properties of steel used in nuclear power plant pressure vessels, he could consult an editor who is a nuclear engineer and/or one who is a mechanical engineer.

Broadening of Ei's Coverage of Technical Literature

The Ei technical translator not only translates literature, but also is involved in editing, abstracting, indexing, evaluating, reviewing and editorial decision-making. For instance, the Editorial Division evaluates new English and non-English publications for addition to the Ei databases. Such new publications are either received directly by Ei or are borrowed from ESL, with which Ei maintains a good and close working relationship. These publications are evaluated for inclusion in the Ei databases by the appropriate editors and supervisors. Needless to say, the number of new technical and scientific publications appearing nowadays is quite staggering and, consequently, only those publications that offer top technical quality and editorial content, and are highly valuable and pertinent to the engineering community are selected for Ei coverage.

The Ei editors are also familiar with the impact and requirements of today's computer-oriented information industry by participating in in-house workshops concerning online retrieval methods of the Ei databases, engaging in an active interface with the other divisions of Ei and attending various online conferences and information-industry exhibits. The need for constant education stems from the fact that Ei continues to expand its already large database. Over 2.5 million abstracts appear in the printed products and on microfilm, and over 1 million are in computer-readable form. Part of this expansion includes Ei's latest service made available in July 1982; the *Ei Engineering Meetings* database covering published proceedings of conferences, meetings and symposia. Ei also plans to further expand its range of services to include numerical and computational databases.

Conclusion

It is hoped that an accurate and faithful picture has been painted of how one organization, Engineering Information, Inc., deals with the task of providing competent technical translation and that the described Ei experience, operating procedures and policies will be of interest to other companies, organizations and individuals involved in similar tasks.

Recognizing both the growth and the potential of the information industry it seems quite certain that technical translation will continue to play a vital role in the transfer of technology. Avenues which remain open as to how technical translation could be made more effective, less costly, and faster will be discussed in a future article. For the immediate future, a competent and experienced translator will continue to provide Ei with the best answer to technical translation.

GOVERNMENT INFORMATION:
WILL AN INFORMED PUBLIC BE SACRIFICED
IN THE NAME OF PRIVATE ENTERPRISE?

Wilda B. Newman

ABSTRACT. An analysis is made of the future role of the private sector in the handling of government information. Possible dangers to the general public that could result are discussed.

NTIS* Take Over by Private Sector Stirs Lively Debate.[1] This was the headline of a news article appearing in the February 15, 1982 issue of *Library Journal.* Such a headline may have been chosen to incite or simply to apprise the information community. If it was intended to incite, there does not appear to be a "move into action" by the library community. If it was merely to apprise, then it behooves the information community to investigate the implications of such a move by the government.

NTIS is only one information facility, and changes as to how federally funded information is indexed, controlled, and distributed are not limited to NTIS. More is astir regarding the whole spectrum of government information services.[2] If NTIS is adversely affected, by extension what of the Government Printing Office (GPO), Defense Technical Information Center (DTIC), Library of Congress (LC), National Library of Medicine (NLM), National Agricultural Library (NAL), and others? You can easily identify more.

There are some articles appearing in the professional literature, and one major report[3] has been written. Miriam Braverman[4] addresses the question of "public good" in her article, "From Adam Smith to Ronald Reagan: Public Libraries as a Public Good." To paraphrase Ms. Braverman, "If *all*

Wilda B. Newman is a Librarian at The Johns Hopkins University, Applied Physics Laboratory, Johns Hopkins Road, Laurel, MD 20707. She is a member of the Committee on Information Hang-ups, Washington, D. C. and Chairperson of their Sub-committee on Government Information Services. Ms. Newman has a BS in Business and Management (Univ. of Maryland) and is currently enrolled in a Master's program for Library and Information Science.

---

*National Technical Information Service

---

information can be bought and sold at a profit, and the judgement is made that information is not necessary to the whole society then it is a commodity, subject to the marketplace and belongs in the private sector.''

Another example of the current discussion surrounding information can be found in ''Shifting Boundaries in Information,'' where Anita Schiller[5] addresses the prediction of the demise of the library and states that ''these predictions come at a time when information as a resource has taken on a new economic value, and the information sector of the national economy is seen by some to be its driving force.'' She goes on to say that ''not-for-profit activities, have become primary, for profit activities . . . [and that] today information is increasingly defined as a commodity to be bought and sold.''

No one argues that these boundaries are not shifting nor do librarians and information specialists quibble about ''the primary goal of the Information Industry Association (IIA) . . . to promote the development of private enterprise in the field of information and to gain recognition for information as a commercial product.''[5] Ethics and propriety are the real issues, the rules recognized in our society that support the doctrine of an informed public, and conformity to those established standards will insure the fruition of this ideal.

We concede that there is room and even need for the private, profit-making sector in information, but we see as well significant implications that must be addressed. Such possibilities as a major medical laboratory taking over the information services currently provided by the National Library of Medicine; a chemical or oil company taking over the information services currently provided by the National Agricultural Library; an airplane manufacturer taking over the technical information services currently provided by the Defense Technical Information Center or the National Technical Information Services; or a food product corporation taking over the familial information services currently provided by the Government Printing Office require serious and immediate attention.

Besides these concerns there are basic questions regarding availability, cost, copyright, Freedom of Information, services, impartial interface, and low-bid contracting that must be dealt with.

Availability

Will aggressive marketing of some technical reports escalate their sales and distribution over equally valid but ''less popular'' reports, i.e., energy reports? Would there be a split in terms of availability relative to those reports that are lucrative and those that are not? How would the ''user community'' know which were which? Who would decide and based on what criteria? Will ''low-demand'' reports become ''low-availability''? or worse, not available at all due to prohibitive costs for storage and maintenance? What will happen with ''out-of-print'' items? Distribution, control, and maintenance of technical reports is much different than that of books. An out-of-print book can generally be borrowed from another library.

Cost

What is the cost to the government if it must buy information it created? Is that cost less than the government handling the information services to disseminate that same information?

Currently small research companies have reasonable access to R & D efforts. If the price of information to them was increased significantly, how would this affect their ability to compete in R & D?

Copyright

What of the issue of copyright? Current programs from NTIS (SRIM) and DTIC (ADD) distribute microfiche copies of technical reports that can be announced and copied (fiche to fiche, or fiche to paper) by the library. Will private industry be prompted to copyright these materials in order to stop such practices in the name of commercial enterprise?

Services

What happens to the current services supplied by NTIS to such agencies as the Department of Energy and the National Aeronautics and Space Administration? What are the questions, considerations and implications for the government agencies such facilities serve? Have the agencies been asked for their opinion on how such a change would effect them and their services as well as their mission. What happens to the service, run by the private sector, if it goes bankrupt or for some other reason cannot honor its commitment to the user community?

Freedom of Information

What of the ''Freedom of Information'' requests to agencies? Can or should a business enterprise fill such requests?

Impartial Interface

How would the impartial interface be continued, between both the agencies and the information services, and foreign countries and information services?

Low-bid Contracting

Low bid can and often does mean low-level service. How is this to be controlled and corrected? There are instances on record where companies giving very poor service continued to win the bid on a contract because they were the lowest bidder and the users of such services had no recourse. What of the problem where a low bid is made to get the account and additional funds are used when needed from another contract (a common practice in some areas)?

While these categories do not include every aspect of concern or cover all the questions librarians and information specialists may have, they do serve to some extent as an indicator of the scope of the problem. The question of whether or not government operated technical R & D information services should become private sector enterprises is not simply a question of economics.

Both the Special Libraries Association and the American Library Association have published statements concerning the changes currently under discussion relative to the government information services, and these statements are important and should carry a considerable amount of weight. However, it is also necessary for us to move individually into action, and immediately as professionals and as *citizens*.

It is simplistic and stereotypical thinking by the current administration that enacting more cutbacks in federal aid to these programs in order to bolster the private sector will in any way benefit the public. On the contrary, it is more apt to jeopardize the public's "rightful access to government information . . ." [It appears that more concern is being given to profit taking and short-term political considerations] ". . .than with the fair and equitable dissemination of information."[1]

". . .IBM's Chief Scientist Lewis Branscomb. . .[states] that the electronic libraries of the future will have in them only the information someone knows we want to know. That's probably not the information we most need to know."[5] The same could be said concerning the future of our government information services, only worse—they may provide only information deemed "profitable" and that the "haves" could afford.

Reducing the size of the Federal Government and boosting the private sector may be viewed as commendable goals. However, it is unreasonable to reduce information to a "fee not free" basis to realize such goals. Smaller government is not necessarily cheaper or better government, and the intended boost to the private sector may simply be a boost to one end while further restricting another, i.e., the technological and R & D communities, both critically important to the national interest, specifically the productivity and economy of the country.

The library community has been accused of non-objectivity because of their vested interest. That is not the issue—control and manipulation of information is the issue. Our vested interest is as "informed" citizens, first and foremost.

The question is not one of pro- or anti-government, nor profit versus nonprofit, but rather a question of pro- versus anti-information and whether one of the greatest democratic strengths of our country, an informed public, is to continue being supported. That should mean a public supported and informed equally.

BIBLIOGRAPHY

1. NTIS takeover by private sector stirs lively debate. *Library Journal*. 107(4): 382; 1982 Feb. 15.

2. Industry to Feds: keep out of databases. *Bulletin of the American Society for Information Science*. 8(4): 6; 1982 April. *Also:* Brown, Rep. George E., Jr. Restricting information: national security versus rights of citizens. *Bulletin of the American Society for Information Science*. 8(4): 36,35; 1982 Apr.

3. National Commission on Libraries and Information Science. *Public sector/private sector interaction in providing information services*. Washington: GPO; 1982.

4. Braverman, Miriam. From Adam Smith to Ronald Reagan: public libraries as a public good. *Library Journal*. 107(4): 397–401; 1982 Feb. 15.

5. Schiller, Anita. Shifting boundaries in information. *Library Journal*. 106(7): 705–709; 1981 Apr. 1.

ADDITIONAL REFERENCES

Werner, Gloria. Editorial. *Bulletin of the Medical Library Association*. 70(2): 244–245; 1982 Apr.

Herner, Saul. Private correspondence to Paul Zurkowski, President, Information Industry Association. 1982 May 14.

Office of Management and Budget. Office of Federal Procurement Policy. *Proposed revisions to OMB Circular A-76*. (Draft). Washington; 1982 March 30.

Office of Management and Budget. *Managing Federal information resources*. First annual report under the Paperwork Reduction Act of 1980. Washington; 1982 Apr. 1; Chap. 4, pp. 25–36.

NEW REFERENCE WORKS IN
SCIENCE AND TECHNOLOGY

Janice W. Bain, Editor

Reviewers for this issue are: Janice W. Bain, Transportation Research Board (JB); Carmela Carbone, Engineering Societies Library (CC); David A. Tyckoson, Science Library, Miami University, Oxford, Ohio (DT); and Barbara Walcott, Health Sciences Library, Columbia University (BW).

CLIMATOLOGY

Weather of U.S. cities. Detroit: Gale; 1981. 2v. 1169p. $68.00 ISBN 0-8103-1034-1.

Climatological data for 293 U.S. cities and weather observation stations is presented in these two volumes, corresponding to the "first order" field stations of the National Weather Service. For each entry (some cities have more than one), there is a short narrative describing the general weather patterns of the area and six statistical tables including temperature, precipitation, snowfall, degree days, etc. Averages are given for each month from 1939–1978 along with the record highs and lows. All of the information included is compiled from the series known as *Local Climatological Data*, so any researcher dealing with a specific locale may wish to subscribe to this series for more comprehensive and up to date information. However, for reference purposes, this compilation is very valuable for anyone interested in climatology and should be included in any science library collection. (DT)

Janice W. Bain is Program Review Coordinator at the Transportation Research Board, National Research Council of the National Academy of Science, 2101 Constitution Avenue, Washington, DC 20418.

GEOLOGY

Carmichael, Robert S., ed. *Handbook of physical properties of rocks*. Boca Raton, FL: CRC Press; 1982. 2v. Price not given. ISBN 0-8493-0226-9 (v.1); 0-8493-227-7 (v.2).

Physical properties of rocks and of their constituent minerals are of concern to geologists, geophysicists, petrophysicists, and geotechnical engineers. They are of interest for recently developing topics such as deeper drilling for petroleum and other resources; understanding and prediction of earthquakes; engineering geology; more refined geophysical prospecting of the subsurface; and study of surface geology from satellite remote sensing. This handbook is interdisciplinary in content and approach. It provides data for researchers and practitioners in geology, geophysics, geochemistry, petrophysics, materials science, or geotechnical engineering who might be expert in one special topic but who seek information on materials and properties in another topic. The format is primarily tabular for easy reference and comparability. In addition to tables and listings, there are graphs and descriptions where appropriate. The properties covered in Volume I include mineral composition of rocks and electrical properties of rocks and minerals. Volume II includes seismic velocities, magnetic properties of minerals and rocks, and engineering properties of rock. A projected Volume III will include mechanical properties (inelastic), elastic constants, thermal properties, seismic attenuation, and radioactivity properties. (CC)

Luttrell, Gwendolyn W. et al. *Lexicon of geologic names of the United States for 1968–1975*. Washington, D.C.: Government Printing Office; 1981. 342p. $7.50. (Geological Survey Bulletin no. 1520).

The newest volume of this standard reference work has finally been completed after six years in compilation. It provides a description of all new and proposed formation names described between 1968 and 1975, along with a few unrecorded earlier names. Each entry includes a description of the formation as well as a reference to the first or most complete citation to it. Other information provided includes location, lithology, age, thickness, color, and neighboring formations. References are pulled from U.S. Geological Survey publications, state geological survey publications, and several geological journals and books. This tool and its previous volumes are essential for all libraries with a collection in geology. (DT)

Smith, David G., ed. *The Cambridge encyclopedia of earth sciences*. New York: Crown; 1981. 496p. $35.00. ISBN 0-517-54370-2.

Written in textbook rather than dictionary format, *The Cambridge Encyclopedia* covers the field of geology in six main categories: historical, physical and chemical, crustal processes, surface processes, earth resources, and extraterrestrial geology. Each part is broken down into numerous articles by experts in the fields. The articles are written for beginning students or the general reader and tend to be nontechnical. With its extensive index, it can be used as either a reference book or an introductory text. The illustrations, including color photos and maps, are excellent. The encyclopedia is a very good introduction to the earth sciences and should be a part of all science library collections. (DT)

Swanson, Roger W. et al. *Geologic names of the United States through 1975*. Washington, D.C.: Government Printing Office: 1981. 643p. $9.00. (Geological Survey Bulletin no. 1535)

This book is a computer-produced list of geologic names commonly in use in the United States as of 1975. For each entry, brief information such as location, age, lithology, thickness, etc. is given. For many entries, a reference is given to the appropriate volume of the *Lexicon of Geologic Names of the United States* for a more complete description. The list is intended primarily for U.S.G.S. geologists using the online database, but it is very useful for a quick verification and reference to a formation. Because of the six-year delay in compilation, it will not be updated frequently, but additions and changes will be published annually in the U.S.G.S. *Bulletin*. This list should become a standard reference tool in all geology collections and would be useful in all science libraries. (DT)

HEALTH SCIENCES

Handicapping conditions and services directory. Detroit: Grand River Books; 1981. 236p. $32.00 ISBN 0-8103-0995-5.

Originally published by the U.S. Department of Health, Education and Welfare's Office for Handicapped Individuals as the second edition of *Directory of national information sources on handicapping conditions and related services* (Washington, D.C.: 1980) this version of the directory, like its predecessor, was compiled to enable the Clearinghouse on the Handicapped to make effective referrals. It identifies 285 organizations providing information and services to the handicapped. Most of the organizations included are oriented to the physically handicapped. The organizations listed range from those that focus on one handicapping condition to large government agencies like the National Park Service that have offices to handle inquiries on services for the handicapped in their areas of responsibility. An abstract describes each organization and the information services it offers. Entries are grouped by type, but there is an alphabetical list of all of them and a good subject index. Relevant online data bases and their vendors are also described. (BW)

MATERIALS

Chaney, J. F.; Ramdas, V.; Rodriguez, C. R.; Wu, M. H., editors. *Thermophysical properties research literature retrieval guide 1900–1980.* New York: IFI/Plenum Data Co.: 1982. 7v. $99.29/v. ISBN 0-306-67221-9 (v.1).

In conjunction with its research activities, the Center for Information and Numerical Data Analysis and Synthesis (CINDAS) screens the world's literature and collects published information on a wide range of materials in the field of thermophysics. This information concerns data, theoretical estimation methods, and experimental measurement techniques. In 1967, CINDAS published a work entitled *Thermophysical properties research literature retrieval guide.* This three-volume work listed 33,700 references on seven thermophysical property groups and about 45,000 materials. This Basic Edition systematically covered the world's unclassified literature published essentially between 1920 and mid-1964. Supplement I— covering the years mid-1964 to 1970—was published in 1973 and in 1979 CINDAS published Supplement II covering 1971 to 1977. In order to incorporate a number of improvements and editorial changes and to avoid repetitive searches by the user in three publications, it was decided to merge the 1967 Basic Edition and the two supplements, add new citations, and update the new work to 1980. This new Basic Edition thus covers the world literature on fourteen thermophysical properties for the period 1900 to 1980. It consists of over 4800 pages comprising seven volumes, each covering a specialized or related group of materials, as follows: (1) elements; (2) inorganic compounds; (3) organic compounds and polymeric materials; (4) alloys, intermetallic compounds and Cermets; (5) oxide mixtures and minerals; (6) mixtures and solutions; and (7) coatings, systems, composites, foods, animal and vegetable products. This should be a welcome reference work to those not acquainted with the earlier versions and a more convenient and complete work for those using the earlier work and supplements. (CC)

OCCUPATIONAL HEALTH AND SAFETY

Ives, Jane H. *International occupational safety and health resource catalogue*. New York: Praeger; 1981. 311 p. $35.95. ISBN 0-03-060299-8.

A conference on the export of hazardous industries from the United States to developing countries gave rise to this publication. It is in four parts: a list of organizations, a short bibliography, a series of appendixes, and indexes. In the first section, organizations with an expressed interest in environmental quality, hazard export, or occupational safety and health are listed geographically. Most entries consist of a description of the organization and a brief bibliography of its publications. The appendixes are a substantial portion of the book and provide supplementary information such as addresses of labor education programs, general information on OSHA activities, and a list of occupational safety and health materials produced by labor unions. In spite of its rather awkward arrangement, this should be a useful resource on a subject of increasing importance. (BW)

SCIENCE

Bynum, W. F. et al, eds. *Dictionary of the history of science*. Princeton, NJ: Princeton University Press; 1981. 494p. $40.00 ISBN 0-691-08287-1.

New dictionaries abound in the sciences, but this is one that is truly unique. It is the first to define science in historical and philosophical terms rather than the standard definitions used in most dictionaries. Each definition presents the basic development of a concept as well as the thought changes behind that development. The entries range from a few lines to pages, with bibliographies in the longer ones. Many cross references are given, so that a concept can be traced through many entries. An eleven-page bibliography on the history of science is also included. This dictionary is recommended for any library collection concerned with the history of science. (DT)

SCIENCE & TECHNOLOGY

Gibson, Robert W. Jr.; Kunkel, Barbara K. *Japanese scientific and technical literature: a subject guide*. Westport, CT: Greenwood Press; 1981. 560p. $75.00. ISBN 0-313-22929-5.

This subject guide is the result of a study by a special project group in the General Motors Research Laboratories Library to attempt to define the dimensions of Japanese research documentation. The study focused on the following questions: (1) What is the size and scope of Japan's science and technology documentation effort? (2) How much of the Japanese science and technology literature is accessible in the West? (3) Does the accessible portion of this literature document the ''best'' research being done in Japan? and (4) What problems does the language barrier pose for Western researchers on the areas of accessing and using Japanese publications? Part I of the book is an analysis of information activities and bibliographic control in Japan for scientific and technical literature. Part II is the subject guide covering 9,116 titles in science and technology published in Japan. A modified Universal Decimal Classification scheme was used to categorize each title by subject. In this system, the natural sciences correspond to the pure sciences used in Western classification schemes, and the applied sciences correspond to the area known as technology in Western schemes. The titles in this guide are categorized by the modified UDC scheme and under each classification the titles are arranged alphabetically. Each entry gives the information in the following order: (1) romanized Japanese title, (2) title in English, (3) publisher and/or editing body in romanized Japanese, (4) English translation of publisher and/or editing body, (5) address of the publisher, (6) language of publication, and (7) codes indicating abstracting and indexing services which index each title. (CC)

TRANSPORTATION

N. D. Lea Transportation Research Corporation. *Dictionary of public transport*. Washington, D.C.: 1982. n.p. $25.00. ISBN 3-87094-783-7 (available from GAM Printers, P.O. Box 353, Sterling, VA 22170).

This tri-lingual dictionary—in English, French, and German—contains some 2000 public transportation and related terms and their definitions. The compilers' goal is to encourage standardized use of transit terminology in the international transit community. The dictionary was sponsored by federal transportation agencies in the United States, Canada, and the Federal Republic of Germany. The International Union of Public Transport (UIPT) has endorsed the project. It is anticipated that the dictionary will be Part I of a forthcoming *International Transit Handbook and Compendium*. The dictionary is divided into three sections: by language with each term—in alphabetical order—showing the equivalent term(s) in the other languages. The English-and German-language sections give brief definitions for each listed term; the French section inexplicably does not (although it is assumed that definitions for the French-language section will be included in the forthcoming *International Transit Handbook and Compendium*). Where an English-language term is used primarily in the United Kingdom or in North America, it is so indicated by the use, respectively, of "UK" or "AM." Where the definition uses related terms, such terms are preceded by an arrow and enclosed with quotation marks. These terms appear in their appropriate alphabetical sequence. Where a transit term in one language has no equivalent term in another, or is used commonly in other languages, the term is included in alphabetical sequence in the other section(s) of the dictionary. (JB)

SCI-TECH ONLINE

Ellen Nagle, Editor

Database News

Biotechnology Database Announced

TELEGEN, a new file from the Environment Information Center, is being offered by DIALOG as File 238. *TELEGEN* provides information related to the fields of genetic engineering and biotechnology. Scientific, technical and socioeconomic information is derived from over 7000 worldwide sources and includes references to conference and symposia papers, government studies, professional journal articles, news stories, university and corporate reports, and independent laboratory studies dating back to 1973. The database corresponds to the annual *Telegen Reporter Review* and is updated from the *Telegen Reporter*. As of September 1982 the file contained 5300 records; updates add 300 citations per month. Abstracts are included in records added from March 1982 forward.

The subject coverage of *TELEGEN* includes the following topics: business and economics; patents and legal issues; industrial microbiology; energy and mining applications; environmental applications; pharmaceutical applications; research on human, animal, plant, yeast, fungus, bacterial, and viral genes; social impacts; policy and regulatory issues; chemical applications; biohazards; food processing and production; medical applications. *TELEGEN* costs $90 per connect hour, $.15 per full record printed online, and $.25 per full record printed offline.

PAPERCHEM Access Enhanced

SDC has announced price reductions and improved search features for *PAPERCHEM*, the database of the Institute of Paper Chemistry. In addition, DIALOG is now offering the database on its search system. *PAPER-*

CHEM is an international database covering the literature of pulp and paper technology and related topics such as the chemistry of cellulose, hemicellulose, carbohydrates, lignin, and extractives; engineering and process control; corrugated and particle boards; forestry; graphic arts; corrosion; equipment; packaging; pollution; water; and power. Nearly 1000 periodicals are screened as well as the patent gazettes of six major countries. Coverage also includes books, dissertations, symposia, and translations. Approximately 95% of the records include abstracts; a controlled vocabulary is used for indexing. The *PAPERCHEM* file has about 160,00 records from July 1968 to the present. Some 1100 records are added monthly.

SDC's new rates are $55 per hour for non-IPC members (reduced from $110 per hour), and $49 per hour for IPC members (down from $80). Print charges are $.18 per citation online and $.23 for each offline print. To permit browsing, there is no charge for online scanning of documents. SDC is also offering proximity and direct searching of titles, index terms, abstracts and source fields.

DIALOG is now providing *PAPERCHEM* as File 240 for $75 per connect hour. Offline print costs are $.15 per full record.

ASI Available from DIALOG

The *American Statistics Index (ASI)* is being offered as File 102. Produced by the Congressional Information Service, *ASI* is a comprehensive guide to the statistical publications of the U.S. federal government. It covers the publications of more than 500 sources within the executive, legislative, and judicial branches of the government. Every type of federal government publication is indexed and described in *ASI*, if it contains statistics.

The subject areas covered by *ASI* reflect the total range of interests of U.S. government agencies: agriculture, medicine, health, education, sociology, economics, demography, natural resources, and the environment. Housing and industrial reports, international trade data, labor statistics, and sources of statistics on crime, taxes, finance and banking, transportation, and energy are also included.

Records on *ASI* include full bibliographic information on the source, an abstract describing the publication's subject matter and purpose, and an outline of specific contents. The *ASI* database contains approximately 100,000 records from 1973 to the present. About 1000 records are added monthly. The cost is $90 per connect hour and $.25 for each record printed offline.

National Library of Medicine Completes CANCERLIT Regeneration

In July 1982 a regenerated *CANCERLIT* database was made available to MEDLARS users. The regenerated file contains more than 306,000 records. The most important change made to *CANCERLIT* during the regeneration was the expansion of subject searching capabilities through the addition of MeSH (*Medical Subject Headings*) vocabulary and special cancer "enrichment" terms. MeSH terms have been added to 98.5% of all records added to *CANCERLIT* since January 1980. Indexers will always index, in depth, the cancer-related section of an article whether it is the major point of the article or not. In the latter part of 1979, a list of terms was developed by National Cancer Institute indexers for cancer concepts that were not covered by MeSH terms. There are currently 49 of these enrichment terms being used.

Bureau of National Affairs Databased Added By DIALOG

DIALOG is offering two new databases produced by the Bureau of National Affairs (BNA) which should be of considerable interest to science and technology libraries: *PATLAW* and *LABORLAW*.

PATLAW is the online version of the *United States Patent Quarterly* which covers the sources of intellectual property decisions of the U.S. Supreme Court, the U.S. Court of Appeals, the District courts of the U.S., the U.S. Court of Customs and Patent Appeals, the Commissioner of Patents and Trademarks, the U.S. International Trade Commission, and selected decisions from U.S. state courts. It contains descriptions of reported judicial and administrative decisions pertaining to patents, trademarks copyrights, and unfair competition law.

Records in the *PATLAW* database include a reference to the case name, the headnote (or abstract) of the case, subject descriptors, and, where applicable, references to the case history and/or paralleled citations. BNA descriptor codes are also available for searching on broad categories. *PATLAW* contains approximately 51,000 records from 1970 to the present; it will be updated monthly with about 250 records. The price for searching this database (File 243) is $120 per connect hour and $.70 for each record printed offline.

LABORLAW (File 244) provides summaries of decisions and references to the source of the text of decisions on matters relating to labor relations, fair employment, wages and hours, and occupational safety and health. The database is comprised of six major subfiles covering the following: *Labor*

Relations which includes abstracts of federal and state decisions involving labor and management and National Labor Relations Board rulings back to 1966; *Labor Arbitration Reports* including references from 1969 forward covering labor dispute settlements; *Fair Employment Practice Cases* with information from 1938 to the present on federal and state court cases and Equal Employment Opportunity Commission rulings; *Wage and Hour Cases* covering all such cases back to 1961; *Occupational Safety and Health Cases* with information from 1972 forward covering Occupational Safety and Health Review Commission decisions as well as federal and state court decisions; *Mine Safety and Health Cases* which contains information covering decisions of the Federal Mine Safety and Health Review Commission, and key federal court decisions. *LABORLAW* contains approximately 175,000 records. The file is updated monthly with 1500 records. File 244 costs $120 per connect hour and $.70 per offline print.

CAS ONLINE Adds To Search Services

The *CAS ONLINE* search service now displays full bibliographic information for the ten most recent references for a substance retrieved in a search. It will also be possible to place orders online for the referenced documents. Orders will be processed within 24 hours; telefacsimile transmission will also be available.

SDC Offers TSCA

SDC recently announced that the *Toxic Substances Control Act (TSCA)* database is now available on the ORBIT system. TSCA is produced by the Office of Pesticides and Toxic Substances of the U.S. Environmental Protection Agency (EPA). By law, all substances produced, processed, or imported (unless specifically excluded) must be reported to the EPA and included in the database. According to SDC, *TSCA* is valuable to searchers for three reasons: it is a comprehensive list of all chemicals commercially available in the U.S.; it links a substance to the producer, and in some cases gives a contact name; and it contains production details by substance and producer.

The *TSCA* database is two databases in one: the Chemical File Segment, the online equivalent of the *TSCA Initial Inventory* and the *Cumulative Supplement II*; and the Plant File Segment. The Chemical File Segment contains about 58,000 records on commercial substances including details of who produced or imported the substance, in what quantity, and whether the substance was used by the plant itself or sold commercially. The Plant File Segment contains information on more than 3000 manufacturers, including name, address, contact name, EPA region, and Dun & Bradstreet number. All substances in the Chemical File Segment include CAS registry number, molecular formula, chemical name, and synonyms. Several fields are proximity searchable.

Charges for *TSCA* are $45 per connect hour and $.15 per citation for off-line prints.

Population Index Added To POPLINE

The National Library of Medicine has announced the addition of citations from *Population Index*, the quarterly bibliographic journal published by the Office of Population Research, Princeton University, to *POPLINE*. More than 15,000 citations taken from issues of *Population Index* published from 1978 to the present are included. Citations from 1981 forward include abstracts.

Population Index was first published in 1935. It is an annotated bibliographic journal which provides comprehensive coverage of demographic literature with particular emphasis on Western and Slavic language publications. Specific subject coverage includes demographic methods and theories, population size and growth, spatial distribution, mortality, fertility, nuptiality, family size, migration, historical demography, population characteristics, population policy, official vital statistics, and population censuses. The work of collecting and abstracting bibliographic material for *Population Index* is funded by the Center for Population Research, National Institute of Child Health and Human Development, National Institutes of Health. The support for the contribution to *POPLINE* comes from the United States Agency for International Development.

Although there is some overlap between the subjects covered by *Population Index*, the Population Information Program of Johns Hopkins University, and the Center for Population and Family Health/Information Program of Columbia University, a thorough selection process is being followed to avoid duplication of records on *POPLINE*, according to NLM.

Abstracts Added to ISMEC

Users of *ISMEC*(*I*nformation *S*ervices in *ME*Chanical Engineering) will find that beginning with the January 1982 update, abstracts have been included in over 90% of *ISMEC* records. Since 1973, the *ISMEC* database has served as an index to the world literature in mechanical engineering. Abstracts are currently for display only, they are not searchable. *ISMEC* is available from DIALOG as File 14.

SCI-TECH IN REVIEW

Suzanne Fedunok, Editor

Full Text Serials On-Line

The availability of full text serials databases is on the increase, as reported in the literature. INFORM is making available the full text of two Philadelphia newspapers, and expects to add others. G. Cam of France is offering full text of *Who's who in Europe*, and Wiley announced full-text availability of the *Harvard Business Review*. (Reported by E.H. Levine. Business and full-text data top Online agenda. *ASIS Bulletin* 8 (5):31; 1982 June). A new company called Comtex Scientific is developing the first of what it hopes will be a line of 22 electronic journals in the sciences. The story of this new venture, and of the reactions of scientists to it, is reported by W.J. Broad in Journals: Fearing the electronic future. *Science* 216 (4549): 964–966; 1982 May 28.

Outlook on Science and Technology

Susan C. Perry. *A report on selected issues in science and technology and their long-range implications*. National Academy of Sciences, Washington, D.C. An $110,000 grant for 12 months: PRM 82-06308.

The Committee on Science, Engineering and Public Policy (COSEPUP), consisting of representatives of the councils of the National Academy of Sciences, the National Academy of Engineering, and the Institute of Medicine, has received funds to prepare a background report on issues in science and technology and their long-range implications. The report, designed as an annual follow-up to the successful National Research council *Five-Year Outlook* series, is to be called *Annual Science and Technology Report and Outlook*, and is required by the National Science and Technology Policy, Organization and Priorities Act of 1976 (Public Law 94–282). The body of the report, due to be submitted to the Congress in February 1983, will consist of a series of individually authored chapters, each discussing one area of recent advance in science and technology and each authored by an eminent scientist or engineer. The report will also include an overview and a chapter on applications, to be written by COSEPUP members acting as a body.

Electronic Mail Analysis

Electronic mail. Report No. E65. 107 p. Predicasts, Inc. 11001 Cedar Ave., Cleveland, OH 44106.

Predicasts, Inc., has just completed an analysis of the electronic mail industry which predicts that there will be a 12-fold increase in electronic mail messages by 1995, and that most of the growth will be concentrated in intracompany communications. Businesses are most likely to be affected by this trend, according to Predicasts, because 80% of first class mail is business related. They predict that 30% of the first class mail market will be captured by electronic mail systems by 1995, mostly because of new product developments, decreasing equipment costs, increasing equipment standardization, and developments in communications technology such as digitalization of the telephone network and satellite and terrestrial broadband networks using microwave and optical fibers.

Inventory of Software Packages

National Center of Scientific and Technological Information (COSTI), P.O. Box 20125, Tel-Aviv 61201 Israel.

The Israeli National Center of Scientific and Technological Information (COSTI) has prepared for UNESCO an inventory of software packages compiled from over 150 responses to a detailed questionnaire designed to evaluate software packages for mainframe, mini- and microcomputers. The questionnaire sought information on hardware requirements, application packages, database management systems, transferability, availability to developing nations, users's experience, costs, performance, etc.

The inventory describes software packages which serve or may serve in textual or alphanumeric information work, library and documentation systems, SDI and online searches, information and fact storage, retrieval and distribution, text processing and publication. Only those systems which have been in operation successfully for some time are included, and the inventory gives tables based on the questionnaires that are to enable users to compare systems and to identify those which come closest to their needs.

Freedom of Information Act Indexes

Guide to freedom of information indexes. *Information Hotline* 14(7): 9–12; 1982 July-August.

The Freedom of Information Act requires federal government agencies to make available for public inspection and copying current indexes identifying for the public information on any matter issued, adopted, or promulgated after July 4, 1967. A guide, including a number of indexes from the National Science Foundation among others of interest to science and technology libraries, has been compiled by the Office of the Federal Register from information submitted by agencies from January to September 1981 and issued in tabular form. Headings of the guide include: Issuing agency, Subagency name, Index title, Period covered, Brief description of the contents, "Order from" information including price, Copying, and Additional information including a contact person's name. Examples of entries are an NSF index of publications by topic, NSF grant policies and procedures, and NSF memoranda from the Director and Deputy Director.

New Publications on Computers

Chen, C.-C. and Schweizer, S. *Online bibliographic searching: a learning manual.* Neal-Schuman Publishers, 23 Cornelia St., New York, NY 10014. $19.95.

Ching-Chih Chen and Suzanne Schweizer have just published a new guide to bibliographic searching designed for classroom use or independent study which features a section on the future on online searching.

Davis, Charles and Lundeen, Gerald. *Illustrative computer programming for libraries.* Greenwood Press, 88 Post Road West, Westport, CT 06881. $15.

Another new book features examples of how to do programming for libraries, and includes samples of programs that can be modified to suit local library conditions. User-oriented programs such as document retrieval are explained, as are numerical problems such as circulation tallies and the number and cost of acquisitions, examples of keyword indexing and corcordance preparation. It is well illustrated and has numerous examples.

Computing Newsletter, Box 7345, Colorado Springs, CO 80933.

The 1981 fifteenth edition of the *Annual bibliography of computer-oriented books*, compiled by the University of Colorado, has appeared. It contains more than 1000 books from over 170 publishers, and has been revised to exclude all introductory books published prior to 1979. The books are classified according to type (reference, text, handbook) and arranged in 63 subject categories with notes on the type of presentation (narrative, programmed instruction, case study). The best thing about the bibliography perhaps is the price: $4.

Directory of Data Sources

Among the new publications available from commercial publishers now is the *CODATA Bulletin*, which is publishing serially a *Directory of data sources for science and technology*. The following chapters are now available from Pergamon Press at $10 each:

> Chapter 5: Seismology. (CODATA Bulletin No. 42, June 1981)
> Chapter 6: Chemical kinetics. (CODATA Bulletin No. 43, July 1981)
> Chapter 7: Nuclear and elementary particle physics. (CODATA Bulletin No. 48, June 1981)

Technical Reports From NTIS

Patents and Trademarks

The Patent and Trademark Office has issued the following series of machine-readable data files, which may be useful to science and technology libraries in compiling their own databases:

PB81-208050　*Patent full text file.* (PTO/DF-81/002)
> This file contains full text information on U.S. patents issued since 1970, although it is not a complete list for entries prior to 1974, because the storage of patent information on tape was being tested and developed during that period. The data file includes patent number, patent classifications, inventor's name, and patent titles. It does not

include drawings or molecular configurations of chemical patents.

PB81-208100 *Sequential classification title file.* (PTO/DF-81/007)
A comprehensive listing of the U.S. Patent Classification System, updated regularly (usually semiannually) to reflect changes in response
to new technology.

PB81-208084 *Concordance file.* (PTO/DF-81/005)
This concordance relates the U.S. Patent Classification System to the
International Patent Classification (IPC), as published for the World
Intellectual Property Organization by Carl Heymans Verlag in Munich.
The difference in the philosophies between these two systems unfortunately means that this concordance can only serve as a guide to lead
the user to a group of related adjacent categories of classifications, even
in those cases where a one-to-one correspondence between entries
seems to be indicated.

PB81-208118 *Index to Classification.* (PTO/DF-81/008)
A list in alphabetical order by subject headings and subclasses as
assigned by the U.S. Patent Classification System is given. It is intended
to be of use to those unfamiliar with either the classification system
or with the particular technology to be looked up.

All of these data files can be transferred onto most standard 7 or 9 track
recording modes for one half inch tape. The source tape is in EBCDIC
character set, and the requestor should identify the recording mode desired
by specifying: character set, track, density, and parity. For price and further
information, contact: NTIS Computer Products, National Technical Information Service, P.O. Box 1553, Springfield, VA 22161

Microfiche

This document reports results of a study by the National Institute of Justice
on the cost effectiveness of disseminating reports on fiche instead of paper
copy. The conclusion reached is that most of the benefits of fiche accrue
to the producer and not to the user. The report predicts great difficulty in
accustoming this user group of criminal justice professionals to the use of
fiche, but it does offer several suggestions, including emphasizing the role
of libraries, of using fiche in new ways, instead of as simple one-to-one
substitutes for paper documents. PB81-228009 *Microfiche as a vehicle for
technical reports.* (PC A03/MF A01)

A Non-Success Story

The refreshing viewpoint of this memorandum report by Michael Stonebraker of the University of California at Berkeley's Electronics Research Laboratory is to look at the mistakes that happened in the implementation of the INGRES database system, rather than to report on the final corrected version. The role of structure design and the problem of non-trivial users were critical to the project and are discussed in some detail. The report closes with some "miscellaneous impressions" of UNIX, PDP-11 and data models that might prove useful to librarians.

AD-A103 115/2 *Requium for a database system*. (PC A03/MF A01)

Microforms

This 271 page document represents a bibliography of 263 citations that update earlier bibliographies of 1979 and 1980 on the management of microform collections, including microform format standards and the utilization and care of microforms. It includes 52 new entries not found in the earlier bibliographies.

PB81-80687 *Microforms, 1973–1981. Citations from the NTIS database*. (PC N01/MF N01)

Libraries

This 63 page report is actually the proceedings of the Department of Defence Longe Range ADP Planning Conference, held in Long Beach, California, in January 1981. Selected senior Defence Department and other agency information technology managers heard presentations of 21 panelists speaking on five topics: information resource management, trend projections and ADP policies for the 80s. Program management, and office automation. These are abstracts (1000–2500 words long) of these presentations.

AD A099 441/8 *Managing information technology change in the 80's*. (PC A04/MF A01)

The American Library Association and Mary Jo Lynch undertook this 236 page handbook to provide all types of libraries—academic, public, school, with the tools for collecting basic management information that must be gathered in all kinds of libraries to meet administrative and planning objectives. The handbook provides basic definitions of the important categories of information to be gathered and then supplied methods for its collection, with hints on how to tailor the collection to the library involved.

PB81-203184 *Library data collection handbook*. (PC A11/MF A01)

Recreational reading from NTIS

This bibliography compiled by the NTIS contains 118 citations to literature on video games of skill, both fixed programming and freely programmable. The hardware and software, including circuitry, microprocessors, and video modules, are considered. The emphasis is on home programmable video games. The citations come from the International Information Service for the Physics and Engineering Communities Data Base.

PB81-874968 *Video games for home and commercial use. January 1975–August 1981*. (PC N01/MF N01)

NSF Mathematics Research Institutes

In 1981 the National Science Foundation funded for a five year period two mathematical research institutes, patterned somewhat after the Institute for Advanced Study at Princeton. The pure mathematics institute is at Berkeley, and the other one is at the University of Minnesota. The Minnesota institute will focus each year on a broad topic of interest to applied mathematicians in solving real-life problems. The first topic will be "statistical and continuum approaches to phase transition" and the second year's topic will be "mathematical models for the economics of decentralized resource allocations." Publications will probably emerge from the Institute's activities that will be of interest to many researchers and librarians.

The Mathematics Library of the University of Minnesota is just completing a project sponsored by a Mellon grant and the Research Libraries Group to do a retrospective conversion of its entire monographic holdings records into machine-readable form for RLIN (Research Libraries Information Network). This effort will result in the one of the largest collections (15,000 volumes) of monographs in mathematics becoming part of the RLIN database, and represents one of RLG's first efforts to do a retrospective conversion.

Marge Voelker, "Mathematics workshop." *P-A-M Physics Astronomy Math Division on Bulletin of the Special Libraries Association*. 10(1): 14; July/August 1982.

Document Delivery and Reference in Mathematics

Along with the arrival of *Mathematical Reviews* online in the new database *Mathfile*, comes the news from the University of Illinois at Urbana-Champaign Mathematics Library that they have received a Title II-C grant from the U.S. Office of Education that will in effect establish the Mathematics Library at Illinois at the "library of last resort" for items reviewed in

Mathematical Reviews

To quote from the grant proposal:

> This grant will enable the University of Illinois to establish a document delivery and reference center in mathematics at the same time as the American Mathematical Society makes available its online database Mathfile. The 2000 unowned monograph titles reviewed in *Mathematical Reviews* will be acquired on microfilm and added to an existing collection of 35,000 monograph titles and 1300 serial titles. Data from all titles will be entered into the OCLC database, using AACR-II cataloging and latest class numbers. Thus, this project will provide nationwide access to a major research collection in mathematics, by means of standardized, uniform cataloging in the country's largest automated database. *For Further Information Contact*: Nancy D. Anderson, Mathematics Librarian, University of Illinois Mathematics Library, 1409 Green St., Urbana, IL 61801 (217-333-2884).

SCI-TECH NOTES

ONLINE PATENT SEARCHING
AT THE NEW YORK PUBLIC LIBRARY

Happy recent developments in the Patent Section of The Research Libraries of the New York Public Library occasion this update to an article entitled "Patent and Trademark Collection of the New York Public Library" published in this periodical, Volume 2, number 2, pages 55–60, Winter 1981.

The U.S. Patent and Trademark Office has now developed automated tools to help patent examiners and the public search through their files. One of these tools is the Classification and Search Support Information System, CASSIS. This is a computerized index to the classified patent search file and is now available in the Patent Section of The Research Libraries, as well as in other Patent Depository Libraries throughout the United States.

CASSIS presently provides four capabilities or modes:

1. Display of all U.S. patents assigned in a classification.
2. Display of the original and cross reference classification of a U.S. patent.
3. Display of structured classification titles.
4. Search of key words in classification titles.

Included along with the regular U.S. patents are reissues, design patents, and plant (botanical) patents. Unfortunately, as of this writing, trademark information is not available via CASSIS.

A Decawriter, model LA120 (Digital Equipment Corporation, Maynard, Massachusetts), was leased and installed. Public service began on April 30, 1982; through May 18 (fourteen public service days) the system was utilized for 42 searches. In all instances only modes 1 and 2 (as stated above were used. As members of the patent lawyer community become more aware of this service, not only will the number of searches increase dramatically but modes 3 and 4 may also receive their share of inquiries.

For the present only staff members use the system directly to answer inquiries and there is no charge to the public for the service. Staff and public involved are very satisfied (even excited) with this operation and are looking forward eagerly to continued upgrading of the service as bugs in the program are eliminated and availability is advertized.

Robert G. Krupp
Richard L. Hill

For Product Safety Concerns and Information please contact our EU
representative GPSR@taylorandfrancis.com
Taylor & Francis Verlag GmbH, Kaufingerstraße 24, 80331 München, Germany